Corydon L. Ford

OUTLINES OF THE FIRST COURSE

OF

YALE AGRICULTURAL LECTURES.

BY

HENRY S. OLCOTT.

WITH AN INTRODUCTION BY

JOHN A. PORTER,

PROFESSOR OF ORGANIC CHEMISTRY AT YALE COLLEGE.

NEW YORK:
C. M. SAXTON, BARKER AND CO.,
No. 25 PARK ROW,
1860.

EDWARD O. JENKINS, PRINTER, 26 FRANKFORT ST., NEW YORK.

OUTLINES

OF THE

FIRST COURSE

OF

YALE AGRICULTURAL LECTURES.

BY

HENRY S. OLCOTT.

WITH AN INTRODUCTION BY

JOHN A. PORTER,

PROFESSOR OF ORGANIC CHEMISTRY AT YALE COLLEGE.

NEW YORK:

C. M. SAXTON, BARKER & CO.,

25 PARK ROW.

1860.

EDWARD O. JENKINS,
Printer & Stereotyper,
No. 26 Frankfort Street.

PUBLISHERS' PREFACE.

THESE sketches of the lectures which were given during the recent Convention at Yale College, were first printed in the columns of the New York Tribune. Occurring as it did when there was an unusual pressure upon the columns of the paper, the Convention would never have been reported at all if the editors had not regarded with great favor this attempt to improve the condition of our Agricultural science. Anxious to lend the powerful aid of the Tribune to further the object in view, they allotted a sufficient space daily for a succinct outline of the lectures throughout the entire course. So much valuable information was embraced in the several discourses, that to the reporter it was a matter of great difficulty to select as little as would fill the space at his disposal; and the readers of this pamphlet will not, therefore, wonder if he has not done full justice to either the topics or the lectures. When the course was almost completed frequent inquiries were made as to whether any complete report of it would be published; and by many a desire was expressed that if nothing more detailed and elaborate could be done, at least these Tribune sketches should be collected in book form, for convenience of preservation. It being established beyond doubt that no full publication could, for various reasons, be made, the publishers of this volume have made arrangements with Mr. Olcott to edit and correct his notes. To render them as nearly perfect as their brevity permits, they have been submitted for revision to the lecturers themselves, and may, therefore, be considered as at least fair summaries of the matter delivered by them from the lecture-desk.

LECTURES

GIVEN DURING THE

AGRICULTURAL CONVENTION AT NEW HAVEN,

FEBRUARY, 1860.

FIRST WEEK.—AGRICULTURAL CHEMISTRY, &c.

AGRICULTURAL CHEMISTRY,............................PROF. S. W. JOHNSON.

LECTURE 1. Composition of the Plant. The Organic Elements—Oxygen, Nitrogen, Hydrogen, and Carbon. LEC. 2. Proximate Organic Principles of the Plant—Cellulose, Starch, Dextrine, Sugar, Gluten, Albumen, Casein, Vegetable Oils, and Acids. LEC. 3. Atmospheric Food of Plants—Water, Carbonic Acid, Ammonia, and Nitric Acid. Their sources and supply. LEC. 4. The Ash of Plants—Potash, Soda, Lime, Magnesia, Oxyd of Iron, Oxyd of Manganese, Chlorine, Sulphur, Phosphorus.

ENTOMOLOGY,..DR. ASA FITCH.

LECTURE 1. Great losses sustained from depredating insects—their classification, structure, metamorphoses, habits, and means of destruction. LEC. 2. Insects injurious to grain crops, with a particular account of the wheat midge and Hessian fly. LEC. 8. Insects injurious to fruit-trees, with a particular account of the Curculio and the Apple-Tree Borer.

VEGETABLE PHYSIOLOGY,..............................DANIEL C. EATON, ESQ.

LECTURE 1. The vegetable cell—its form, size, structure, contents, origin, and mode of growth. LEC. 2. The seed, root, and stem. Nature and growth of seeds. Structure of roots. General structure and minute anatomy of stems. LEC. 3. Arrangement of leaves—their parts, forms, structure, and economy. Food of plants. Relations of the vegetable kingdom. LEC. 4. Flowers and Fruits. Arrangement of Flowers—their parts and offices of parts ; development of fruit.

VEGETABLE PATHOLOGY,............................CHAUNCEY E. GOODRICH.

SECOND WEEK.—POMOLOGY, &c.

PEAR CULTURE,....................................HON. MARSHALL P. WILDER.

American Pomology—the best method of promoting it ; with practical suggestions on the cultivation of the pear.

GRAPES,..DR. C. W. GRANT.

LECTURE 1. Preparation of the soil, and propagation of the vine. LEC. 2. Culture of native varieties, with an account of different varieties and their qualities. LEC. 3. Foreign varieties ; culture and treatment.

FOURTH WEEK.—DOMESTIC ANIMALS.

INTRODUCTION.

BY PROFESSOR JOHN A. PORTER.

THE views of Agricultural Education in which the Course of Lectures originated—reports of which are here presented to the public—were set forth in the *New Englander*, for November, 1859. From that article we make a few quotations, as introductory to a sketch of the course itself, and of the advantages which may be expected from a pursuance of this system of agricultural education:

"There is little question in the public mind as to the importance of new agencies for the diffusion of agricultural knowledge. A more difficult question is, how the lack of them shall be supplied. The Press does much, but by no means all that is required. The contact of man with man, and of mind with mind, is necessary to inspire the enthusiasm which is essential to rapid progress.

"The introduction of books on elementary science into our Common Schools, would be a great step in advance ; but here again there is the absence of that contact of the man of knowledge with the men who need it, which is essential to the highest success.

"Shall we wait for the establishment by Government of great agricultural institutions, similar to those of continental Europe? Such institutions are among the most obvious and essential wants of our time, but a public and general opinion of their utility and necessity must be created before either our

State or National Governments will seriously consider their establishment. Shall we await the results of private enterprise or beneficence in the creation of agricultural institutions, with their model farms and costly apparatus of instruction, and their corps of professors, exclusively devoted to the business of instruction? For these also we should have long to wait, not so much because of the want of liberality among those who have the means to endow such institutions, as for the lack of a clear conviction as yet of their utility, and the really practical character of the information they would supply.

"It has seemed to us that this problem of a more perfect diffusion of knowledge on agricultural subjects, is capable of another solution than that which consists in devising means for obtaining governmental appropriations, or awaiting the munificence of individuals.

"In the attempts which have hitherto been made in this direction, too exclusive reliance has been imposed, as it seems to us, on purely professional instruction; and it has been wrongly assumed that it is necessary to await the gradual production of a class of men qualified to impart it. No necessity exists, as we believe, to await the creation or production of anything that does not now exist, for the accomplishment of this great work. The material is at hand. We have undiffused knowledge among us in every department of agriculture and horticulture, and of science applied to cultivation, as minute and profound as exists anywhere on the face of the earth.

"In accordance with this view, the solution which we propose is *the enlistment of practical men, who are not professional teachers, in the work of instruction, and their combination in such numbers, that a small contribution of time and labor from each shall make a sufficient aggregate to meet the object in view.* The special necessity for such a system, in the case of the pursuit we are considering, grows out of the fact that there is much in agriculture which has not, as yet, taken the form of Science, and can only be acquired from practical men.

"We are all familiar with the immense results accomplished by combinations of capital in commercial enterprise, in banking, in railroad projects, in manufacturing. The combination which is practicable in agriculture is of another kind—the association of intelligence and knowledge in the work of instruction, for the indirect attainment of great results in this most important of all fields of human labor.

"To realize such association of knowledge we would, then, assemble from the farm, the garden, the nursery, the vineyard, and from the ranks of science, gentlemen distinguished for their skill in the various specialties of agriculture—practical and theoretic,—and call on them to make each his contribution to the work of instruction. And then we would summon the intelligent and enterprising farmers of the country, young and old, to gather and learn from the most highly qualified among their own number, the secrets of their success. We would propose that such aggregations of knowledge, as have been suggested, should be made at as many different points in the country as the available material would warrant, and that the instruction they would furnish should be adapted as exactly as possible, in time and extent, to the circumstances of our agricultural population.

"Such gatherings would partake of the character of the agricultural convention, on the one hand, in which experienced cultivators meet for their mutual enlightenment; and on the other hand, of the agricultural college, within whose walls the less experienced assemble to take advantage of the deliberations of the former, and to listen also to their formal instruction."

The experiment proposed as above, in November last, has since been made under the auspices of the Yale Scientific School. Before proceeding with our sketch of it, a few words may be appropriate with regard to the Institution which has undertaken to carry out this scheme of Agricultural Education.

The Yale Scientific School, is the Scientific Department of Yale College, sustaining the same relation to the parent institution as the schools of Law, Medicine, and Divinity. Its Faculty consists of seven Professors of the following branches, viz.:—Civil Engineering, Industrial Physics and Mechanics, Geology and Mineralogy, Metallurgy, General and Applied Chemistry, Organic Chemistry and Agricultural Chemistry. Its course of study extends through two years. The Engineering Department has recently instituted a third year's course of higher studies, and the new degree of Civil Engineer. Within a few months the school will enter upon the occupancy of a new and commodious building recently erected for its accommodation, at an expense of forty thousand dollars, by a friend of the Institution. This building contains, beside its laboratories, recitation rooms and lecture halls, ample accommodations for an extensive agricultural museum. A handsome fund for this especial object is already accumulated, and will be largely increased and in part expended during the present summer. This movement, although subserving completely the objects of the winter course on Agriculture, has by no means exclusive reference to this course, but is to be regarded as a development of the permanent Agricultural Department of the Institution which remains in session during the whole year.

The new building not being completed as was anticipated, the late course was given in a public hall in the city of New Haven. The lectures were commenced on the first day of February, and on the twenty-fourth day of that month were brought to a close. Twenty-six gentlemen, distinguished in various specialties of agriculture, participated directly in the work of instruction, and not less than five hundred persons were attracted to the city of New Haven during its progress. Three or four lectures were given each day, and the time not thus occupied was devoted to inquiries on the part of the audience, and to discussions thus suggested. These discussions, in which other gentlemen of experience besides the lecturers took an active part, proved to be the most valuable part of the proceedings of this convention.

We proceed to a few remarks, suggested by the experience of the late course, as to the kind and degree of benefit which may be expected from similar conventions in the future.

In the first place, it is obvious that their usefulness is to be found by no means exclusively or even principally, in the novelties in agricultural science or practice which are likely to be presented in the lectures. Every important discovery in agriculture finds its way, of necessity, into the agricultural journals, and through the newspaper press becomes the property of the country. In addition to this, every important subject on agriculture or horticulture is presented in books especially devoted to the purpose, which the cultivator may study at his leisure without the necessity of leaving his home. These facts might seem at first view to do away with all necessity for such gatherings. They do not influence us, however, to hesitate in the least in declaring them among the most efficient means in existence for promoting agricultural progress.

On the benefits to the experienced cultivator it is unnecessary to dwell. Agricultural, horticultural, and stock-breeding conventions have come to be common and popular, and it is already established by experience that they subserve many important purposes which are unattainable by other means. The statement of numerous individual experiences in such a convention, will frequently show in an hour on which side the balance of testimony lies, and so decide in a brief session questions which have been the subject of a newspaper war of months. A brisk fire of questions will often annihilate, in a few minutes, the carefully guarded statement which has served as the protection of some cherished error, and so expose, by a single attack, the fatally weak point of some plausible theory, which might have been perpetuated in print for years.

Often, also, out of a chaos of seemingly inconsistent testimony there will crystallize by the aggregation of individual experiences a really valuable result, which would never have been attained but by the free interchange of opinions, which is only possible when men meet face to face.

Of the advantage of such conventions to the comparatively inexperienced cultivator, we shall speak with somewhat more of detail.

In the first place, attendance upon them necessitates the absolute and undisturbed appropriation of a certain definite time to the acquisition of agricultural knowledge. At home the time would not have been found; at the convention it is secured. The young farmer who is at the trouble and expense of going abroad for a month for the purpose of study, feels that it is his sole business for the session to learn, as it is on the farm to work. This consideration is an argument of itself almost sufficient for such gatherings. A convention of merest tyros in agriculture, without teachers to instruct or guide them, would of itself be a valuable institution, if only for the definite allotment of time to the business of study. Assembled with such advantages, of instruction the time secured for such an object ensures the most important results.

A second advantage of such conventions is the influence of the living teacher. This, in the case of persons who are without the mental discipline furnished by a course of severe study, is an advantage which cannot well be over-estimated. The young man who will gape in the chimney-corner over an agricultural volume, will listen with intense interest to the very same matter from the lips of an earnest speaker. And the enthusiasm of the teacher will infuse a permanent vitality into the principles he communicates, which will make them living and efficient agencies in the mind of the pupil, instead of mere dead acts accumulated and laid away for a future use which is never realized.

To illustrate by a particular case, we venture to say that the four lectures on Drainage, given by Judge French, during the recent course, did more to make an impression on the minds of the young farmers who heard them, and more to ensure attention to this important means of agricultural improvement, than all the essays on the subject which they had ever perused. And the same principle might be illustrated by many other lectures of the course.

A third advantage of such conventions is to be found in the illustration of the subjects presented by specimens and experiments, by drawings and models, and by living plants and animals. This is an incalculable advantage which the private library and the home study cannot furnish, and which places this mode of instruction for definiteness of information immeasurably above all others. Mr. Barry whittling at his pear-tree before the audience, is worth a whole treatise on grafting and pruning. Mr. Gold's discourse on sheep, interspersed with the bleatings of his Cotswolds, and punctuated with the black noses of his Southdowns, is worth a volume on mutton and wool.

Still another advantage of such gatherings is to be found in the opportunity they afford to the pupil of eliciting from his instructors knowledge especially adapted to his own particular case. Books are dumb to such inquiries, and even the elaborate treatise often leaves unnoticed the particular point which is essential, in order to give the rest value for any particular locality. It is for this reason, as before stated, that the inquiries, replies, and discussions which are regarded as essential features of this method of education, are also its most efficient agencies of instruction. These are by no means confined to the lecture-room. During such a convention every hotel and boarding-house is the locality of an agricultural club, which is in session during the whole of the twenty-four hours not devoted to the public meetings and to sleep.

Finally, we remark, that the mere contact with men of great experience and high success in agriculture, is stimulating and inspiring to the young agriculturist as no mere shadow of their personality in print can possibly be. They stand before him as living illustrations of the great results of fortune and of reputation which may be achieved by energy and enterprise in this noble field of labor. They encourage him also by the impression which their personal presence will not fail to make, that these results are not a consequence of great intellectual superiority, of freedom from doubts and difficulties, and of mysterious insight into the processes of nature, but of quiet and persistent labor, to which he also is equal, of science which he can attain and of enterprise which he himself can rival.

If any one has been disposed to inquire whether the newspaper reports of the proceedings of such a convention do not furnish a large part of the advantage which would be derived from attending its lectures and deliberations, the reply which we are disposed to make to such an inquiry will already have been inferred. While serving perfectly its purpose of giving to the public a general idea of the proceedings of an Agricultural Convention, the newspaper can furnish at best, consistently with its other offices, but a small fraction of the matter of the mere lectures of such a course. Should it furnish all, it would supply but the mere skeleton of their value to which the life and blood of inquiry and discussion and special application, and the electricity of personal influence and enthusiasm, would be wanting. Detailed reports, which should record the total proceedings, including inquiries, replies, and discussions, are out of the question, from the space they would occupy and the expense they would involve. But if practicable, they would be destitute of all the peculiar advantages which have been rehearsed as belonging to the system. These are to be found, if we may be allowed here to recapitulate, in the appropriation of a definite period to the work of study, in the substitution of oral for written instruction, in the facilities afforded for special inquiries, in the opportunities furnished of obtaining valuable knowledge in private conversation, in the personal influence of the instructor, in the intercourse with eminent cultivators, and in the complete illustration to the eye of every subject which is presented to the mind.

In relation to the present reports, although they are far from needing any apology, it is but justice to the reporter to say that they were made during the hurry of a convention, six to eight hours of whose time were occupied every day with public meetings, and under a pressure of material which compelled him to make selection his object, rather than completeness. In justice to the lecturers, it is proper to say they are not to be held responsible for any inaccuracies of statement which may possibly have crept into the reports, or for the occasional inadequate presentation of their discourses. This was often necessitated by the pressure of other matter on the columns of the paper for which the reports were prepared. A few omissions which occurred, from the same cause, have been supplied from other journals, at the suggestion of the writer of this introduction. One of the gentlemen who took part in the course, regarding it as entirely impracticable to give brief reports any practical value, has requested that his lecture should be omitted in this publication. His wishes have been respected by the publishers.

Let the enterprising farmer, who would attach his sons to the calling to which he has devoted his own life, and put them on the road to success in their pursuit, beware of the false economy which is disposed to reason that an agricultural paper once a week, or a report of a convention once a year, is all that is necessary to effect this important object. Let him give his children the advantage of association with the men whose example dignifies and elevates his calling, and demonstrates it as noble a road to fortune and to happiness as any that nature or art has opened. Let him insure for them, by contact with such men, somewhat of the zeal and enthusiasm and knowledge which has been the secret of their success, and the efficient instrument of their advancement. Thus only can so important an object be realized.

Let it not be imagined that in this attempt to set forth some of the advantages of the system of Agricultural Education here presented, there is the least design to depreciate any one of the manifold agencies in operation for the accomplishment of the same great object of agricultural improvement. Of these, perhaps the Press is the most important, and the one with whose influence we could least afford to dispense in the promotion of this cause. But the Press scatters material a large part of which is lost, for the want of leading principles in the minds of its readers which such a system would best furnish, and according to which its countless facts might be arranged. The nucleus of knowledge and enthusiasm once created by such a method of instruction, it would attach to itself these floating fragments of experience and observation, and, like the growing crystal, build them up into its own substance, and make them part of its own life.

The Farmers' Club is a most efficient agency, but it is often a dead and cumbrous heap for want of the fire which might be kindled from such a flame. The Agricultural Fair is a most potent instrument of progress, but, without some system of agricultural education behind it, is a mere confusing chaos of illustrations, comparatively worthless, as the chemist's experiments would be without his explanations, for lack of the knowledge of the great principles to be illustrated. All of these agencies have contributed to make possible the introduction of such a system of agricultural education as is here discussed. The system once in operation would react upon these earlier agencies, and give them increased vigor and efficiency.

The Convention and course of lectures recently concluded, was so far successful as to justify the announcement of its repetion in February, 1861. It is regarded, however, as important chiefly as having furnished the means of determining how such a course may be made most useful and attractive in the future. While retaining, therefore, in the Course of '61, the fundamental idea of this system of Agricultural Education, viz., that of the combined College and Convention, the second course will be carried out with various modifications which have been suggested by the experience which has now been obtained. It will be entered upon with vastly increased means of success, in buildings, collections, and other apparatus of instruction, and also in the wide spread interest which the past course has awakened. Undertaken with such advantages, it will be of especial interest as determining, once for all, the practicability of sustaining such a course of instruction. To this end, an amount of patronage at least two-fold, and probably three-fold that which the late course obtained, is essential, even on the basis of extremely moderate compensation to the lecturers. Whether this can be secured our experiment of next winter will determine.

THE

YALE AGRICULTURAL

LECTURES.

FIRST DAY.—Feb. 1, 1860.

While the friends of an improved agriculture have been for many years advocating this or that reform, and to this day are dolefully wailing over the torpid state of farm science, and praying that something might be done to popularize it, Professor Porter of Yale College, with admirable boldness, has conceived and commenced this first course of Agricultural Lectures at Yale College. He very wisely thought that the man of knowledge should be brought in direct contact with the men who need it, the skilled farmer come face to face with the unskilled, and that, by choosing a number of men, eminent in the several branches of agriculture, to succeed each other in a course of lectures, our farmers' sons, by sparing a fortnight or month in winter, and coming to one central point, would get more information of value to themselves than if they pored over books for a whole year. He plainly saw that if we were to wait for such Governmental aid and comfort to Agricultural Colleges as is given in Europe, he and we all might grow grey and die before our hopes were half realized ; and no more feasible plan suggesting itself, he bethought himself, to use his own language, of "the enlistment of practical men, who are not professional teachers, in the work of instruction, and their combination in such numbers that a small contribution of time and labor from each shall make a sufficient aggregate to meet the object in view." You will understand, then, that mainly to Professor Porter, and not to Yale College, the honor of originating this plan is due. Yale has done something for scientific agriculture since about the year 1848, when a Professorship was partially endowed for the late Prof. John Pitkin Norton, who had labored some time with Johnston in England. Norton died in 1852, much regretted, after having done as much as he could to make his department useful and popular, and was immediately succeeded by Prof. J. A. Porter, who was called from Brown University. Porter's incumbency lasted five years, when he accepted the Chair of Organic Chemistry, resigning his own place to a rising young man, Samuel W. Johnson. Mr. Johnson had studied two years in the Scientific School here, and then went to Germany, where he worked in Leipsic a year, in Erdmann's laboratory, and an equal time with the great Liebig, at Munich, beside making visits to various laboratories and schools in Germany, England, and elsewhere. Since he took his Chair at Yale, he has held the office of Chemist to the State Agricultural Society, and made some notable analyses of muck and phosphates, the latter of which have occasioned much controversy.

He opened the course this morning with an elementary lecture on agriculture, confining his remarks to the organic elements of the plant, and explaining their nature and properties by the usual experiments.

Three lectures are to be given daily (except Saturdays and Mondays, when there will be but two,) until the 25th of this month. The morning lecture is at 11; the afternoon, at 3, and the evening one at 7 o'clock.

The 3 o'clock lecture to-day was by Mr. DANIEL C. EATON, an amateur botanist of this city, who has, I am told, a very extensive herbarium, and has given many years of study to his specialty. His lecture to-day treated of the vegetable cell—

its form, size, structure, contents, origin, and mode of growth. The vegetable cell, he says, is a closed vessel like an egg, and is composed of an outer solid membrane which contains a fluid, and matter floating in the fluid, or attached to the sides. At first the enclosing membrane is very delicate, and is called a *utricle;* if this remains closed throughout its life, it is called "a cell;" if the sides of several adjoining cells disappear, and the series is arranged into a tube, it becomes "a vessel." Cells are the base of all vegetation. The red snow-plant, and the yeast-plant, are single cells. The snow-plant, so graphically described by Kane and other Arctic explorers, is one cell, with little particles floating within. These particles become cells themselves, in time, and the outer coat bursting, lets them escape to commence an individual existence themselves. Cells vary in form in different plants, and even in the same plant they, by overcrowding here and loosening there, get distorted in shape. In the stems of water-lilies some of the cells are star-shaped, while in the wood of trees they are long and pipe-like. The diameter of cells averages from 1-1200th of an inch, up to 1-250th; but the common puff-ball of our pastures, when broken spirts out a fine brown powder, each particle of which is a cell, or *spore* as it is termed, of infinitesimal diameter.

The membranous wall of cells is of different toughness. In the sea-weed, it is very soft; in ash, hickory, and mahogany, very hard; and in vegetable ivory, harder still. Cell membrane never dissolves in water, but swells. It is called "cellulose," and is composed of oxygen, hydrogen, and nitrogen, chemically written thus: C. 12; O. 10; H. 10. The spaces between the cells of a plant are filled variously:—sometimes with air; in the common red cedar, with minute grains of red aromatic rosin; in sumac, with a thick milky sap; and in other plants, with gums. The contents also of cells vary. The growing cells of some plants, as asparagus, are more nutritious, because they contain some nitrogen, which goes toward making muscle in the animal body. A granular matter, a viscid fluid, sap (which is almost water but contains sugar sometimes,) and the green leaf-color, known as chlorophyll, are also contained in the cell. Starch, too, is sometimes there, and each grain of it is organized, and so organized for each plant that the source of a specimen of starch may often be revealed by microscopic examination. Potatoes store up starch in enormous quantities for the use of the next year's seed-ball; but we, thieves that we are, carry off storehouse, contents and all for our own use.

In cells there are acids sometimes; malic is made by the apple, citric by the lemon, and other kinds by others. Starch is insoluble in water, and cannot, therefore, circulate through the plant; but sugar can, and dextrine, which is in its nature somewhat intermediate between sugar and starch. There are two grand divisions in the plant world—the flowering and the flowerless. The former have elongated cells, as well as short ones, but the simpler of the latter class have not. The distinction is not now recognized as universal, although it has been until recently.

I learn that a friend to Yale College is about to make it a magnificent donation in the shape of a building for its Scientific School. The main building is about fifty feet square, and has two wings of equal dimensions, in one of which is to be the Agricultural Museum, in the other a fine laboratory. The first and second floors of the main building are assigned to the Engineering School, the third to a lecture hall.

SECOND DAY.—FEB. 2, 1860.

Dr. ASA FITCH, of New York, gave last evening his lecture on "Economical Entomology," or injurious insects. The Temple, where this convention sits, was about half filled, and the lecturer was frequently applauded. Dr. Fitch labors in a field of science vastly important to farmers, but very poorly understood. As he very justly remarked last evening, the devastations by insects are not noticed, because so insidiously made, but if our eyes could but be opened to the activity of our little foes, consternation would seize us. Go into our forests and we see every portion of our trees attacked by some insect—trunk, bark, leaves, and roots, all having their peculiar depredators. The sweeping away of our forests compels the insects which formerly fed upon them to turn to the orchards, which have replaced the forests. Thus we have the apple-tree borer, which originally subsisted in the wild thorn-apple; and the Buprestis, from the oak; and from present indications it is probable we shall hereafter see the branches of our apple-trees lopped off as are the limbs of the common red oak in particular years, and by the same insect, the "oak-pruner." But in addition to these native species, quite a number of foreign insects have been imported in the thousand commodities, and in the numberless trees and plants which we import, and these have proved the most pernicious foes to our crops and trees. Our crops and climate favoring their development, they multiply to a frightful extent, and do far greater damage here than they did in Europe. The bark louse, for instance, on both sides of Lake Michigan, has ruined nearly every orchard. For years after the settlement of this country wheat was an absolutely sure crop, but the yield dwindled with successive years, and now, in large districts, its culture is necessarily abandoned. Reasons have been urged to account for this; that our soil has deteriorated, and our climate changed, but they do not explain the difficulty. With the best of manuring and tillage, we cannot get the crops our ancestors did with shiftless farming; and even where new woodland is cleared, and wheat is put into the virgin soil, the crop is infinitesimally small. The true cause is to be found in the attacks of insects, and nothing else. The wheat midge and the Hessian fly are the only insects which have attracted much notice, and it is hence currently supposed that these are the only important depredators which we have in our wheat fields. But, a few years since, on coming to examine the growing wheat, the learned lecturer had been surprised to find in every field, multitudes of the Chlorops, Oscinis, and Thrips, insects which have long been known in Europe as most pernicious to the wheat crops there, but which have never been suspected as occurring upon this side of the Atlantic. Some of these depredators are preying upon it at every stage of its growth, the root, the tender blade, the stalk, the ear, and the ripening grain in the ear all having particular enemies infesting them. Now originally, when our country was covered by an unbroken forest, there was no wheat here, nor other plant of the wheat kind, on which such insects could subsist; consequently when the lands were first cleared and sowed to wheat a bountiful harvest was gathered. But the thrifty fields of this grain, with which our country then abounded, invited these insects to them. One after another arriving and finding here an ample supply of its favorite food, would remain, ever afterwards laying the crop under contribution for its support. Thus, as these enemies successively penetrated the country and became established in our wheat fields, their productiveness gradually diminished, till at length it was no longer possible to grow this grain with profit, and in all the older sections of our country its cultivation has long been abandoned. To form some idea of the immense losses these pests are occasioning, look at the wheat midge, which has been ravaging our fields for the past twenty-five years. To appearance it is an insignificant little yellow fly, only a fourth the size of a mosquito; but though it seems so powerless and inert, it was able in New York State, in 1854, to destroy wheat to the value of over $15,000,000, or nearly as much, probably, as the whole city of New Haven is worth, with all its houses, buildings, and lots. If an invading army had destroyed property to this value, how the whole country would have been aroused! Multiply this tremendous loss by that sustained in all the States, and what a result is there for our contemplation! The wheat midge, however, is, sad to say, not our only insect enemy, for the name of the army is legion. And what has rendered the situation of our farmers and fruit-growers most vexatious, they have been obliged to remain in ignorance, no definite information respecting the names and habits of these creatures, from which they are sustaining such losses, being accessible to them. Only two works on this subject have ever appeared, and neither of these has been on sale in the bookstores. One of them is Dr. Harris's Treatise, originally prepared as part of the Natural History Survey of Massachusetts. The other is Dr. Fitch's own Report on Noxious Insects, published each year in the New York Agricultural Society's Transactions, and also issued separately, two volumes being now completed.

The insect is divided into three principal parts, viz.: head, thorax or fore-body, and abdomen or hind-body. The head in insects is furnished with antennæ or horns, which possess remarkable sensitiveness. Thus, an ichneumon fly, by touching them against the outer surface of the bark of a tree in which a worm is lying, detects not merely its presence, but its exact position, although imbedded two or three inches in the solid wood, so accurately that with its long ovipositor or sting it is able to pierce the wood to where the worm lies, and puncture its skin and insert an egg therein. And two bees or ants meeting, by merely touching their horns together, know if they belong to the same hive or hillock—for all the world as though there was a system of Freemasonry among them, whereby they know on this shaking hands as it were, whether they are brothers or strangers to each other.

The most wonderful thing about insects is their metamorphoses, or transformations, the same individual appearing at different times under forms as different as for a serpent to change into an eagle. There are four of these forms or stages in the growth of insects:—first, the egg; second, the larva or growing stage, when it is a worm or caterpillar; third, the pupa or dormant stage, when it is often enclosed in a cocoon; fourth, the perfect insect, when it is a fly, butterfly, beetle, bee, &c.

An insect may be known to be in its perfect or mature state when it has wings; or if it be a wingles variety its maturity is known by its depositing eggs. In grasshoppers, plant-bugs, and leaf-hoppers, the changes are less complete, they never having the form of a worm, the young resembling the mature insect, only being smaller and without wings.

Insects, however much we may despise them, have a real use in the domain of nature—destroying all that is dead, and checking the increase of all that is living in the vegetable world. Without them the earth would immediately be overrun with plant life. And hence those trees and plants which it is man's object to cultivate, come to be attacked by those insects whose office it is to repress these kinds of vegetation. To be successful in his labors, therefore, man is obliged to combat those insects which thus prey upon his crop. To do this he must study their habits and transformations.

Dr. Fitch closed by stating, that the more he examined these creatures, the more confimed he became in the opinion, that there is no injurious insect but that, when we become acquainted with all the details of its history and habits, we shall be able to detect some assailable point and devise some measure by which either the insect can be destroyed or the vegetation can be shielded from its attacks. We shall discover that, although he may be invulnerable in every other part, no ægis protects his heel, and if we strike Achilles there, we inflict a death wound. A prolonged outburst of applause, on the close of the lecture, attested how deeply Dr. Fitch had interested the audience.

Subsequently, in confirmation of Dr. Fitch's statement, that it was not a deterioration of the soil nor change of our climate that prevented our growing such crops of wheat now as formerly, but was the insect enemies of this grain with which the country has become overrun, a gentleman from Maine reported that in a remote part of that State, where a district has recently been newly cleared, distant from where wheat has ever been grown, the finest crops of this grain are now produced. Another gentleman stated, he was satisfied this was also the true solution of a fact that appeared quite singular and unaccountable, viz.: that here on some of the old lands of Connecticut, excellent crops of wheat have recently been grown. The cultivation of this grain had been so long abandoned here, that all these wheat insects have probably disappeared, and thus released from them, these crops that have occasioned so much surprise, have grown on the old lands here, without any special manuring or other management of the crop.

Dr. Fitch lectured again this afternoon, his subject this time being "Insects injurious to Grain Crops, with a Particular Account of the Wheat Midge and Hessian Fly."

He said that our losses are immeasurably greater from insects than those of European nations; as we have not only our own, but many foreign ones introduced here, and these latter often greatly surpass in their destructiveness, with us, anything recorded of them in their native haunts. And yet, because of not being so overcrowded in population, they were not felt so much; for there the loss of one-eighth of a crop would be regarded as a great national disaster, whilst here it would scarcely be noticed.

The Hessian fly was undoubtedly introduced into this country, as at first supposed, in some straw used for package, by the Hessian troops which landed at Flatbush, L. I., August, 1776. The few insects thus brought here multiplied so that in 1779 the wheat fields in that town were destroyed. And from thence it gradually spread in every direction, advancing about twenty miles a year, penetrating to every part of our country. It is a small, white, footless worm, which changes to a pupa resembling a flax-seed, found at the crown of the root in autumn and winter, and at the next June another generation nestles at the lower joints of the stalks. Within a year or two of its first arrival in any given place, most of the surrounding wheat fields were destroyed, and its ravages usually continued for several years, or until its parasitic enemies had multiplied sufficiently to subdue it. It has frequently reappeared here and there, but for many years now, little has been heard of it. This is probably the same insect that is mentioned by Duhamel as having greatly injured the wheat in Switzerland in 1732, and again in 1755; but during the half century of its worst ravages here, it lurked undetected in Europe, till in 1833 it ravaged a part of Germany, and in 1834 was found by Prof. Dana along the Mediterranean in every wheat field he visited in Spain, Italy, and on the Island of Minorca; and finally, in 1852, much damage was caused by it upon the River Volga, where its parasite was also found accompanying it. Such is, in brief, all that is known of the European history of this insect, which, introduced upon our side of the Atlantic, has caused a loss of uncounted millions of dollars.

The wheat midge has long been known in England. It was originally supposed to be a sort of mildew which thus blighted the wheat, and was only ascertained to be an insect in 1771. And in 1797, Mr. Kirby, searching for the Hessian fly, partially traced out the habits of this insect. It was doubtless introduced into this country in some unthreshed wheat brought to Canada, for it was first noticed upon the St. Lawrence, and also in Northern Vermont, in the year 1830, though it did not multiply and become so destructive as to attract public notice until nine years later, when it also began to extend itself, and has now overspread Canada and all the Northern States as far west as into Indiana. Its larva is a minute footless worm, or maggot, of a bright orange-yellow color, found in numbers upon the young kernels in the wheat heads, causing them to be small and shrivelled, to such an extent some years that many fields are not harvested, every kernel being blighted. In England the midge is preyed upon by a parasitic insect, a small kind of ichneumon fly, which rapidly multiplies whenever the midge becomes numerous, and thus quells and subdues it, just as the

Hessian fly with us is now kept in subjection by its parasite. And Dr. Fitch thinks the reason why the midge is so vastly more numerous and destructive here than it ever has been in Europe is, because this parasitic destroyer, its inveterate enemy, has never reached our country. Thus we have received the evil without the remedy. There are two ways by which it is in our power to abate this evil; by destroying, 1st, the fly itsel ; and 2d, its larva. If, early in June, in the evening, when the flies in a swarm are dancing about the wheat heads to deposit their eggs therein, the field be swept over with a suitable kind of net, the flies may be captured therein, and destroyed in such multitudes that the few that are missed will be able to do little injury to the crop. Of the larvæ, a portion remain in the wheat heads at harvest, and are taken into the barn, and are finally gathered among the screenings of the fanning mill, which should be burned, or fed to poultry, and not thrown out, as they usually are, among the litter of the barn yard, where they mature and hatch another swarm of flies. The other portion of these larvæ have at harvest descended to the ground, where they repose slightly under the surface till they hatch into flies the following May; and it has been thought that by plowing the wheat stubble they would be buried so deep as to smother them; but experiments are needed, to demonstrate whether this idea is well founded—these larvæ being very tenacious of life. Water will not drown them. Dr. Fitch has kept them submerged in vials of water three months, and then on placing them on paper they begin to wriggle and crawl away.

The audience being invited to ask questions on the subject of the lecture, if so disposed, availed themselves of the permission. Dr. Fitch, in answer to sundry queries, said that neither sowing lime on wheat when the dew was on, nor sowing salt, nor using sulphur or salt in the granary, nor tobacco-water sprinkled on the field, were specifics. Donald G. Mitchell suggested, as it was uncertain whether deep plowing would destroy the larvæ, the European practice of paring and burning the surface might be resorted to, in the stubble of wheat fields. Dr. Fitch presumed this would be effectual, as the little rascals probably can't stand fire as they do water.

If New York loses fifteen millions of dollars a year from the wheat midge, why wouldn't it be a good plan to send Dr. Fitch to Europe to procure the great foe of the midge, the ichneumon fly? This latter insect sweeps the other from the very face of the earth; and a half-bushel of its eggs hatched on Dr. Fitch's place would be worth its weight in diamonds "of purest ray serene."

THIRD DAY.—Feb. 3, 1860.

Mr. Eaton's lecture on vegetable physiology last evening comprised full descriptions of the seed, root, and stem of plants; the nature and growth of seeds; structure of roots; and the general structure and minute anatomy of stems. He showed, among other things, how the shape of trees is controlled. When the bud at the end of the stem is strongest, the shape of the tree is a pyramid, as in the case of the spruce and fir. Where there is no one strongest terminal bud, there is no principal trunk in the upper part of the tree, so that the tree is rounded at the top, as the elm.

The morning lecture to-day was by Dr. Fitch, and was highly interesting. And here let me state, that, in my opinion, the entomological lectures of Dr. Fitch are the most important of this course, for he shows the habits of, and suggests remedies against, the insects which cause losses to our farmers to a fabulous amount annually; and he stands almost alone in his specialty. The Doctor's lecture to-day was on the insects injurious to fruit-trees. There are at present known to us, in the United States, 60 different insects which prey upon the apple,

12 on the pear, 16 on the peach, 17 on the plum, 35 on the cherry, and 30 on the grape. Prominent among these is the plum weevil, or curculio, which Dr. Fitch stigmatized as the worst insect of our country; for though the midge is at present causing a greater amount of pecuniary loss, he thought its career would be like that of its predecessor, the Hessian fly, and that it would eventually be mastered and subdued by its parasite destroyers. Unlike the wheat midge, the curculio is a native insect of this country, which has now been known upwards of a century, during all of which time it appears to have gradually multiplied and increased its forces, without any important cessations or intervals in its ravages—no parasite destroyer of it having ever been discovered till within a few months past. It was first noticed by the botanists Collinson and Bartram, in 1746, as totally destroying the nectarines in and about Philadelphia, while the plums were but slightly molested. Their turn came next, however, and each subsequent investigator found it ravaging a different section of country. Notwithstanding the volumes written upon it, we do not to this day know where the curculio lives, and what it is doing for three-quarters of the year. All that is currently known of it is, that it is a small brown and white beetle, which makes its appearance on plum-trees when the young fruit is half grown; that it cuts a crescent-shaped slit upon the side of the fruit and drops an egg into the wound, from which egg a small white worm hatches, which burrows in the fruit, causing it to wilt and fall from the tree, whereupon the worm crawls into the ground to repose for two or three weeks during its pupa state; and that it comes out in the latter part of July a beetle, like the parent which six weeks before stung the fruit. This, which is currently supposed to be the main and essential part of its history, Dr. Fitch judges to be quite the reverse; and he is convinced that if there were no fruit for the curculio to eat, it would still thrive to its entire satisfaction.

In New England and New York, the beetle may be found

abroad the last of March, if the weather is fine, though usually it is not till about the middle of May; and in a week or two after it becomes quite common. It is found standing or slowly walking upon the trunk and limbs of the plum, cherry, apple, the wild thorn-apple, the butternut, and other trees. Those on the butternut are plumper than the others. From this time onward, till cold weather returns, we continue to meet with it, and late in autumn it is to be seen on the flowers of the golden-rod as plentifully as at any time through the season. When the young fruit appears, in June, it attacks it with the skill of an epicure, selecting the choicest varieties first. Its crescent-shaped incision is the signal of destruction, as was the crescent banner of the Moslem of old. The slit made, one egg is deposited; and but one slit is made on a fruit. The peach, plum, and apple, when stung, wilt and fall; but the cherry and thorn-apple do not. This is because the larger fruit contains a sufficient amount of nourishment to mature the worm; while the smaller ones must grow on to elaborate the quantity of food which the worm needs. It is a fact not generally known, that apples are attacked by the plum curculio, yet so great are the losses of this particular fruit, that the lecturer gave it as his opinion that the poorer yield of our orchards now, as compared with heretofore, is due to this insect. The wilted fruit literally covers the ground, under many trees, the fore part of July. Cut into this fruit and you will find the same curculio worm therein as in the fallen plums.

From the fact that this insect comes forth three weeks before there is any fruit ready for it to eat, and remains after the fruit is gone, Dr. Fitch thinks that it has other places of refuge to cradle its young besides the young fruit. In fact, it is well ascertained that it breeds in the black knot excrescences on plum and cherry-trees, as eagerly as in young fruit. Hence it has been thought to cause the excrescences. But having examined the black knots fully in every stage of their growth, Dr. Fitch says decidedly they are not produced by this or any other insect, nor are they a vegetable fungus, but are purely a local disease of the limbs, in which the bark and wood are swollen and changed to a spongy substance, but without any of the juiciness which belongs to young fruit. This disease has some analogy to the cancer in the human body, and its cure is the same, namely, the knife, removing the diseased part totally, as soon as discovered.

With Melsheimer, Dr. Fitch believes that the curculio breeds in the bark as well as the fruit of trees, for on a specimen of pear-wood sent him some years ago, his microscope revealed crescent cuts in the bark, like those on young fruit, in which little maggots were lying side by side, ready to eat their way onward when the warmth of spring revived them.

Within six months D. W. Beadle, of St. Catharine's, C. W., has sent the Doctor a curculio parasite, which is furnished with a bristle-like sting with which it pierces the black knot to where the curculio larva lies, and deposits an egg in the body of the latter, to hatch and gradually kill it. The late David Thomas, of Union Springs, New York, first recommended knocking the plum-tree to remove weevils. The remedy is partial, but not infallible. Mr. A. P. Cumings, of New York, recommends to syringe the trees with a mixture of four gallons lime-water, four gallons tobacco-water, one pound whale-oil soap, and four ounces sulphur. The tobacco and soap in solution Dr. Fitch thinks good, but doubts whether the other ingredients add anything to the value of the mixture. There is much testimony to substantiate the fact that trees, whose limbs project over water, always bear fine crops of plums,—the curculio being aware that its young will drown if the fruit drops into the water.

Another important insect is the apple-tree borer,—a long grub which resides under the bark and bores into the solid wood, sometimes below, but usually slightly above the ground, and is two or three years in getting its growth. A few years since, an agent of one of our large nurseries canvassed Wash-

2*

ington county, N. Y., disposing of trees to the amount of three thousand dollars. More than half of these trees have since been destroyed by this borer—a direct loss of $5,000 from this insect in that single county, in addition to the labor lost in planting and nursing these perished trees. This must not be confounded with the borer in the roots of peach-trees, which is the progeny of a moth, while this is the young of a brown, long-horned beetle, having two white stripes the whole length of its back. Specimens of this, as of the other insects spoken of by the lecturer, and of the wood as perforated by it, were passed from hand to hand through the audience. The common soft soap rubbed on the bark of the trees the latter part of May, prevents the attack of this insect. If this be neglected, and the borers have made a lodgement in the bark, their presence is usually shown by particles like sawdust, which they thrust out of their burrows, and when discovered they should be cut out with a knife or chisel without delay.

The regular lecturer of the afternoon was Mr. Eaton, who enlarged on the physiology of vegetables, giving many interesting illustrations of the varied forms and sizes of leaves, and showing how the juices circulate from root to top, and the food is taken and appropriated. He spoke of the essential distinctions between the animal and vegetable kingdoms, and of their relations to each other. Plants are continually purifying the air, rendering it fit for animals to breathe; and plants also, directly or indirectly, supply animals with all their food. Plants live directly on the mineral kingdom, and assimilate to themselves inorganic matter; while animals consume organized matter only.

Mr. Eaton is an enthusiastic botanist, and evidently familiar with his subject.

FOURTH DAY.—Feb. 4, 1860.

A change has been made in our programme. Instead of the third lecture being at seven in the evening, it is transferred to half past three in the afternoon, the usual hour for the second lecture, which, by this arrangement, will be changed to quarter past two o'clock, the two lectures following one after the other. This plan is to accommodate persons who, living out of town, wish to hear all the three lectures, and return home before evening.

Professor Johnson gave a lecture last evening, on the "Atmospheric Food of Plants," reserving a consideration of their inorganic food for this morning. The larger part of the substance of plants is, as every intelligent farmer knows nowadays, obtained from the air; a fact fully proved in the simple experiment of burning wood in our stoves. A log of wood so large as to require two men to roll it on to the fire, burns away so that, after a time, nothing remains but a shovelful of ashes, so light that a child can carry it out. Where has the log gone to, and where have the myriad million tons of trees, plants, and animal bodies gone to, which, in past ages, grew upon the earth? They have each borrowed a little mineral matter from the ground, and a vast quantity of gases from the atmosphere, out of which all their roots, trunks, stems, leaves, and branches have, with wonderful skill, been built. The animal feeding upon the vegetable—it, too, has built up its structure from these same original elements. In both plant and animal the season of life was followed by a time of death, and the organized body resolved into the gases and minerals, the use of which it had borrowed for a brief season. Professor Johnson explained the gradual progress of knowledge of atmospheric constituents, until one day none of its ingredients remained unknown; and by means of the few well-known experiments he demonstrated the nature and properties of each. When the source of the carbon of plants was still a matter of dispute, Boussingault, the great French chemist, proved that only from carbonic acid was it obtained, by the experiment of supplying to a plant under a bell-glass, a weighed quantity of the gas, and noting the proportion abstracted by the plant. The weight of carbon in the soil being absolutely known, as well as that in the plant itself, the increase of quantity at an advanced stage of growth was found to have been attained at the expense of the carbon in the gas, and not of that in the soil.

Mr. Johnson stated it as the practice of some nurserymen to place a piece of carbonate of ammonia, as large as a walnut, upon the steam-pipes of the hothouse. The ammonia thus evaporated produces in the leaves of all the plants with which it comes in contact a splendid deep-green color, and greatly promotes the growth of the plants.

To-day he treated on the ashes of plants, and in the course of his lecture uttered some doctrines which sadly conflict with the received notions which are to be found floating through our agricultural papers. For instance: he said that, chemically, magnesia is *not* injurious to crops when added in excess to the field. The noxious effect of strong magnesian lime, if any, was due simply to a mechanical action in the soil; this particular lime acting in some wise as a cement when moistened. Again: he said that the stiffness of straw is most decidedly *not* owing to an abundance of silica on the outside, but to "the denseness of cellular tissue in the stalk." This he considered proved in the fact that we get from the leaves of the oat and other plants a greater proportion of silica than from the stalk, and yet all leaves are pliant and soft. And the addition of wood-ashes, caustic-lime, and other alkalies with the view to making soluble silicates for the use of the plants, is a piece of useless folly, for "all water found in the soil contains silicates and silica in excess beyond the wants of plants. The addition of alkaline silicates to the soil would be unavailing, for the silicates would be decomposed and the silica rendered insoluble." As an example, he stated that in marshy lands, where sedge and otheraquatic silici ous plants grow, the addition of lime, which removes the excess of silica from the soil, favors the growth of less silicious plants. The silica then on corn-stalks, cereal crops, bamboo, rattan, and such-like, he deems an excretion. The "lodging" of crops he thinks may be owing to a weakness of cellular tissue, which may arise from a lack of some nutritive matter or another, or from excessive transpiration of water. It is known that a plant sucks, sponge-fashion, its juices from the soil, through the extremities of its roots and rootlets. In this water all sorts of mineral matter are dissolved, and with them a certain proportion of carbonic acid and ammonia; well, the plant has a very wonderful power of selecting from this soil-moisture just as much mineral matter as it needs for its growth, and of rejecting all the surplus. Water, however, oozes in, by the principle of endosmose, and is sucked upward from cell surface to cell surface, until it gets to the leaves, where the blowing of wind and the shining of sun upon the leaf surfaces evaporate the water through the little pores, *stomata*, which communicate with the outside air. The plant wants only just so much juice passing through it at once, and if an excess is poured through throughout a warm, damp season, you see how likely it is that its constitution should be weakened. Recent German experiments which have come to Professor Johnson's observation suggest that the beneficial effects of salt, plaster of Paris, and other mineral fertilizers, are due to their preventing this excessive transpiration, or rushing of an excess of water through the plant. Mr. John Johnston sows five bushels of salt on his wheat-fields, "to give stiffness to the straw and prevent rust." The old farmer observed the effect; our chemical friends think they have discovered the cause.

Moreover, what Professor Mapes will scout as sheer heresy, Johnson says that the mineral phosphate from Estramadura and elsewhere is as good for fertilizing crops, if it be properly divided mechanically, as bone phosphate—thus directly combating the Professor's theory of "the progression of primaries by their use in organic nature." Mr. Johnson is a young man, and a bold man; and if he has enough facts to base these several assertions upon, I don't blame him for having the manliness to proclaim them. I must say I like this transpiration theory, for it explains a good many little matters for which a reasonable solution has not heretofore been afforded. As to the stalk-coating affair, and the mineral phosphate business, the case does not as yet seem to me fully proven.

FIFTH DAY.—Feb. 6, 1860.

The Rev. Chauncey E. Goodrich, in his lecture, on Saturday, considered the potato-disease in all its several relations, a branch of investigation on which many years of practice enable him to speak understandingly.

The potato, in a state of nature, is found on the sides of the Andes, and in the adjacent valleys. At the base of the mountains are the tamarind, yam, and banana; the melon, corn, tomato, and pepper come higher up; and above these is the belt where the potato thrives most vigorously, the climate being equable, and the root not exposed to the frosts. When the same varieties of potatoes, especially those which ripen at nearly the same time, are cultivated together, they are variously subject to disease. Thus the old "Early Mountain June," "Early Pink-Eye" or (Dyckman), of the early kinds; and the "Carter," and "Western Red" of the late sorts, are peculiarly liable to disease.

If you plant alongside them, however, the imported "Rough Purple Chili," the "Garnet Chili," the "Black Diamond," and the "Early Hartford," they show a much hardier constitution. And this difference, Mr. Goodrich thinks, is due to a difference in vital energy, which may be owing to a course of replanting, without recourse to the seed-ball, unreasonably protracted. Very wet, cold seasons, such as 1857; or hot, damp ones, like 1850, 1851, and 1855, cause rot; so do sudden alternations of temperature—for instance, from dry, hot weather, to wet, cold, and windy; and these changes destroy the cucumber, squash, melon, tomato, and egg-plant, as well as the potato. The years 1847, 1848, 1854, and 1856, and especially 1852, were favorable ones.

Soil as well as climate has much to do with the nature of crops. Gravel or loamy soils are best, especially when they contain a large proportion of vegetable matter. Sods or straw laid in the furrow over the seed are good, because they maintain an equal temperature beneath them. It is bad to apply much stable manure or guano. Of exposures a northern is best; a southern heats too much, and an eastern heats too rapidly after a cold night. Early planting is best, as it gives the plant a slow, hardy growth in the comparatively wet weather early in the season, which fits it better to withstand the sudden transitions of midsummer.

Early maturing sorts are the surest in bad seasons. Potatoes require deep plowing, and should be subsoiled when a few inches high. Plant six inches deep if your soil be dry, cultivate frequently until the plants are in flower, and *never afterward.* Plant free-growing sorts three by three feet, to give full quantity of air and light. The pieces of seed should not be less than three ounces in weight, each, and cut them lengthwise, never across the potato.

Usual Signs of Disease.—A wilted leaf on the young rosettes of the plant, which are the tenderest parts, and first show disease. 2d. Steel-blue points on some of the older and outer leaves, and yellow iron-rust stains on the inner leaves. 3d. Mildew, which quickly follows these signs, and which, if not arrested, kills the whole plant. These are the signs of disease produced by cold and wet weather changes.

The hot, muggy atmosphere causes an intense dark green color in the leaf, with spotted blotches, which soon turn into mildew, and kill the plant. In the case of cool weather, the flowers fall without setting fruit, while in the hot and damp climate seed-balls set freely, but, with the whole plant, fall a prey to mildew.

The cause of disease Mr. Goodrich believes to be the facility with which a weakened cellular structure will pass into fermentation, in presence of albuminous matter. For a remedy he advises to mow off, or pull up the tops, when it is evident that the weather will not speedily change for the better, but even this will be unavailing in some cases; so that to my mind all this goes to show that our only remedy is to cultivate as well as we know how, choosing new and hardy sorts of potatoes, planting early, and trust to chance for the rest.

The mowing of tops has been tried over and over again, with sometimes success and sometimes the reverse; and so have a thousand other remedies, each of which has in turn been proclaimed a specific. A prize-essay in the Royal Society's Journal for 1858, gives us to understand that deep planting is the true and only remedy; and yet I have planted deep—and so have thousands of others—and yet lost a crop. Mr. Goodrich has spent years in close observation, and accumulated a fund of information, but I venture to say that even he has not yet explained this mysterious disease, its origin and antidotes, so clearly that he who runs may read.

This morning Mr. Eaton spoke briefly about flowers and fruits, showing how the pollen, or yellow dust of the flowers, acts on the ovules or rudimentary seeds; causing them to develop into seeds containing an embryo, and capable of growing up into new plants.

From this he went on to the subject of hybridization, and then of grafting. Grafting has been practically known for many centuries—in fact since the world was young; but the theory was left to botanists to discover. Between the bark and wood are what are called *cambium* layers, or the growing part of the tree, the one which possesses the most active vitality. Unless these cambium layers of the tree and graft are brought together, no union will result; nor will there be one from the contact of very different trees, such as a pear-graft on an oak. The reason for this is, that the cellular tissues of the two are so very different that there is no probability of making a fit, any more than one can fit a sphere to an octahedron. Pears graft well on quince, thornbush, and shadberry. They can be grafted on the apple, but not profitably. The peach goes on to the nectarine, and the plum to the cherry. There are instances of natural grafting, as with the ivy when two branches cross and rub the bark off so as to expose the cambium layers. Of different grafts, of course the best is that which provides for the greatest contact of the layers.

Seeds are of varied vitality. Oily seeds do not keep well because their oleaginous contents are liable to become rancid. Thus the seeds of coffee, magnolia, clove, and such like, must be soon planted or never. Seeds require warmth and moisture, and if kept away from warmth, they often will keep for years and years. Cucumber seeds have been kept seventeen years; corn, thirty; French beans, thirty-three; and from one bag of seeds the Jardin des Plantes was supplied with sensitive plants for sixty years.

To keep seeds well for the longest possible time, gather them when fully ripe, and keep them cool and dry. How wonderful the provisions of Nature for the dispersion of seeds! Some are furnished with feathery wings or silken down, with which they float along on every zephyr; others have barbed points, or hooks, to catch and cling to passing animals; others have elastic capsules or seed-bags, which, when brushed against, burst suddenly apart and scatter the contents abroad; and a thousand other methods might be named, alike curious and admirable.

SIXTH DAY.—Feb. 7, 1860.

Professor Johnson has boldly set himself in array against a new theory of Liebig's, for one thing, and scouts the utility of soil-analysis, for another. Those who have read Liebig's recent pamphlet on "Modern Agriculture," will remember his doctrine that mineral matters are not in a soluble state in the soil; in support of which he quotes the experiment of passing through a sample of fertile soil water holding in solution phosphoric acid and other plant foods, and thereby removing the salts entirely. The formerly soluble mineral matters he supposes to have been made insoluble in the passage through, and putting this and that together, he says that if this be the case, why then, plants must actually have the power of taking in the insoluble material which they need for their growth, and making it soluble after it gets within their spongioles. Johnson thinks Liebig's theory would be very pretty if the little *if* were removed. In other words, he says that Liebig's experiment was rudely performed, and that the mineral matter was not and never can be entirely removed from the water, and hence Liebig's superstructural argument falls, like the Pemberton mills, for want of a sound basis. He says he knows of beans and other plants having been grown and ripened in naught but a watery solution of mineral and organic food—a fact which goes far towards proving that soluble matter is used to full advantage by plants when they can get it. Although I do obeisance to Liebig, I think Johnson is right in this instance, and so I fancy do many others. As to soil-analysis, Johnson reasons thus: One foot deep of the soil in an acre weighs 2,000,000 pounds; a crop of wheat will remove say 200 pounds; if that 200 pounds be not in an available state, no crop will grow. To know if there be enough for the crop, you take a little sample, say 100 or 1,000 grains, and analyse it. Now, does any man living expect the chemist to tell, by even the most miraculously sensitive balances or tests of the infinitesimal sample, whether the 2,000,000 pounds contain enough phosphoric acid, or ammonia, or other ingredients to raise a crop? Take a barren soil, for instance, or one called so, on which the application of 400 pounds of guano will make all the difference of sterility or a crop. Now, can a chemist tell in his laboratory, by testing 100 grains of that soil, taken promiscuously from all parts of the field, whether the guano had or had not been added? Verily not, says Professor Johnson. And so our young agricultural chemist takes issue on the question, and is prepared to do battle with our beautiful pet theory *à l'outrance.* He thinks that if one would take 50 pounds of soil, and wash it with an enormous quantity of water, to dissolve out the soluble salts—a little job which would take at least a fortnight, and might a month—he might, by analysis, find whether there was a great excess or deficiency of plant food in the field from which the sample came. But the cost and trouble of the experiment are serious objections to putting the scheme into practice.

The most fertile soils contain the finest particles; or, in other words, soils are like linen, better for having fine texture. Most soils are deficient mechanically rather than chemically. There is great store of plant food, but not finely enough divided. A field, therefore, which, in a certain state of pulverization, will produce 15 bushels of wheat, would, or should, yield 30 if worked up twice as fine. Why? Because there is twice the amount of surface of particles exposed to the action of heat, and cold, and rain, and therefore twice as much plant food set free. Take your multiplication table and figure up this idea as far as you like, and then you will see the use of sub-soil plows, and clod-crushers, and good harrows, and deep plowing, and all these modern contrivances for breaking up our fields into a good seed-bed.

Last evening the Temple was crowded to hear Mr. WILDER'S excellent address on American Pomology—a topic on which no one in America can speak more understandingly than the President of the National Pomological Congress.

Mr. Wilder commenced by saying that he had accepted the polite invitation of Professor Porter, at considerable inconvenience, for the purpose of bearing his testimony in favor of the present course of lectures. Whatever might be thought by profound scholars of the enterprise, he entertained no doubt that the mass of our practical and intelligent citizens would welcome it as the harbinger of a brighter day in the cause of progressive and general education. The honor of inaugurating this course belongs to gentlemen of Yale College—an institution second to no other in this land for large contributions to the Republic of Letters, for discoveries in the natural sciences, and for their application to the rural arts.

Few subjects exhibit so remarkably the progress of civilization as the increase of fine fruits. In the progress of pomology two facts are worthy of special notice: First, the rapid multiplication of varieties; secondly, the high character of our criterion or standard of excellence. The lecturer here gave a historical account of the progress of fruit-raising, both in Europe and in our own country, mentioning that the first Horticultural Societies in our own land were the Pennsylvanian, and Massachusetts, in 1829, and that of New Haven, in 1830. Now there are more than 1,000 agricultural and horticultural societies, all laboring together, and making pomology a prominent object of support. In 1817 there were no nurseries of any note in New England; now there are many. Then Western New York was just beginning to be settled; now Rochester is the great pomological emporium of our country, and contains the largest commercial nursery in the world. It is estimated that the nurseries of Onondaga and adjoining counties contain fifty millions of trees for sale. Fruit was formerly a luxury; now it is numbered among the common bounties of Providence, and the most humble cottage is rarely without a fruit-tree or a grape-vine.

Our country has taken a leading part in this enterprise. Native fruits are fast superseding foreign varieties. The trees and plants of a country flourish better at home than elsewhere; hence all our efforts are being, and should be, put forth, to get new native sorts of first quality. Of the 36 kinds of apples recommended by the American Pomological Society for general cultivation, 30 are natives; so are 10 out of the 14 plums, half the pears, and all the strawberries. Formerly our only native grapes were the Catawba and Isabella; now they are received in such quantities from the South and West, that a Boston dealar buys two and a half tons at one time for his own trade. A mania now exists for American sorts, some of which will doubtless prove excellent.

A kindred subject is the manufacture of native wine. A Boston manufacturer produces annually, from the wild grapes grown on the banks of Charles river, 20,000 gallons; Connecticut manufactures annually 200,000 gallons; Ohio, 800,000 gallons; and one vine-grower at Los Angelos, Cal., manufactures annnally 2,000 barrels from his own vineyard. Missouri, in addition to her vineyards, has five millions of acres suited to grape culture.

All the strawberries used to be brought from the fields, and not a single American variety had been raised by hybridization; now a cultivator in Massachusetts produces 160 bushels, valued at $1,300 per acre, and another in Connecticut more yet, from new sorts produced from seed. Other parts of the country have improved equally with the East. A Boston apple dealer received last autumn 20,000 barrels of apples from Niagara county, N. Y. In the fall and winter of 1858–9, Boston exported 120,000 barrels, mostly Baldwins. The progress of fruit culture is well illustrated in the returns of the fruit crop of Massachusetts. In 1845 it was valued at $744,000; in 1855 at $1,300,000, and in 1860 it will be $2,000,000, or over.

The soil and climate of the South, contrary to common opinion, are favorable to the culture of fruit. There is an orchard in Georgia of 9,500 pear-trees, and another in Mississippi of 15,000. Many fruits nearly worthless at the North are rendered valuable under the warmer climate and genial sun of the South. One gentleman at the South sends North every year from seven to ten thousand dollars' worth of peaches, before they are ripe in the middle States. We can approximate to an estimate of the fruit crop of the United States from these examples, but who can tell what will be its importance when the numberless young trees planted in the Eastern and Middle States—when the vast vineyards and orchards now flourishing in the great Valley of the Mississippi, and in the Southern States, shall have arrived at maturity?

Col. Wilder next passed to the inquiry, "What are the best means of promoting this art and science?" First, Thorough drainage and the proper preparation of the soil. The former is the great distinguishing feature of the terra-culture of the Nineteenth Century. It is to agriculture what the telegraph and steam are to commerce, and to the progressive civilization of the world. It is an indispensable condition of success in pomology. A pear-tree standing in drained, deep, and thoroughly-worked soil, produced in a single year eight hundred perfect specimens of its fruit, while similar trees, outside the influence of such cultivation, would hardly yield one hundred each, and these of inferior quality. Second, Appropriate soil and location. No tree should be placed where one of the same species had grown and decayed. A treatise which should specify upon scientific principles the particular locality and kind of soil adapted to each species and variety of fruit, would be a desideratum which some one would do well to supply. Third, Climate and meteorological agencies. Climate as well as soil, controlled the quality of our fruit. In cold, wet seasons, fruit was likely to be watery and insipid; in fact, this was so marked as to entirely change the flavor of really luscious varieties of the pear, so that we would scarcely recognize them as the same as we had eaten in propitious seasons. Fourth, Manures and their application.. Analyze your soil and your crop, and manure according to what you find the plant needs. Mulching is an excellent practice. Manure should be applied at or near the surface. An orchard should always be kept free from grass or weeds, and no other crop should be raised except when the trees are small, and even then only a few vegetables midway between the rows. When the trees arrive at maturity, cultivation should not exceed a depth of more than three or four inches; the roots should never be disturbed with the plow or spade. Fifth, The producing *from seed* new and improved varieties suited to each locality. Dr. Van Mons discouraged hybridization. He believed it tended to degeneracy and imperfection, but he must have overlooked the fact that many of his choicest varieties may have been the result of natural impregnation, the pollen being conveyed from one kind to another by the breeze or by insects. Mr. Knight, late President of the London Horticultural Society, was in favor of it. The improvement of plants by this art is illustrated by improvement in the turnip crop of England, of whose importance Daniel Webster remarked: "England would fail to pay the interest of her national debt if turnips were excluded from her culture." But nature's theory is, that like produces like, and the lecturer recommended the planting of the most mature and perfect seed of the most hardy and vigorous sorts. Sixth, The cultivation of the pear upon the quince stock. Some pomologists object to this, but some varieties succeed better on the quince than upon the pear, but they should always be planted upon a luxuriant soil, and be abundantly supplied with nutriment. They should be set deep enough to cover the place where they were grafted three or four inches. In this way the pear would frequently form independent roots, and would combine the early fruiting of the quince with the longevity of the pear. They are well adapted for cities, where garden room is scarce, and for persons advanced in life, who, were they relying on the standard pear for fruit, would die without the sight thereof. Some of the best cultivators practise this plan. The failures in fruit-growing were mainly attributable to bad selection of soil and varieties, injudicious treatment, or bad cultivation. All soils are not suitable for fruit-orchards, nor are all kinds of fruit adapted to every locality. An orchard of half an acre, near Rochester, yielded forty barrels, which sold for $16 per barrel, making $640 for half an acre. Seventh, Pruning, which requires the exercise of the most careful judgment. The pruning-knife of the pomologist is like the amputating knife of the surgeon, to be used only in cases of extreme necessity. As to pruning, it is to be remembered that different varieties require different treatment, for they are not all alike in constitutional vigor, or external form. Hence no general rule could be given; each man must learn from experience. Eighth, Preserving and ripening of fruit. Much progress has been made of late. Fall fruits have been kept till spring. Summer fruits should be gathered before the ripening process commences. The pear, if left to ripen on the tree, forms fibre and farina, but when removed, and placed in a still atmosphere, sugar and juice. Fruits should be kept in a cool, dry, and dark place. About 40° Fahrenheit is the best temperature, but different varieties require different treatment.

The lecturer concluded with a congratulation for those who were entering upon the inviting field of pomological culture. "The innate hope to regain a 'Paradise Lost,' inspires even the most humble to have a country home, and to embellish that home with fruits and flowers. * * * The mission of the pomologist is to multiply our varieties of good fruit—to increase their abundance—to scatter them profusely along the rugged path of life, and thus would he extend the sphere of rational enjoyment, dignify labor, adorn our beloved land with orchards, gardens, and vineyards, and fulfil one of the great purposes of our being—to promote the health and happiness of our fellow-men."

Almost as large an audience assembled this morning to hear LEWIS F. ALLEN speak on fruits. The editor of the American Short-Horn Herd Book showed a familiarity with apples almost equal to that he has with animals, and he gave us his notions in a hearty, good-natured way that enlisted the sympathies of the audience.*

Mr. R. G. PARDEE, of New York city, gave his first lecture this afternoon on the Strawberry. He came, he said, to speak of facts, not theories. He had tried to grow strawberries for many years by high manuring, but without success. He determined to experiment till he should discover the cause of the failure. He had done so. It was by overfeeding. He could now grow them as cheaply as potatoes. The following, according to his experience, is the best method: Select a warm, moist, but exposed situation; for early berries, let it slope to the East or South; for late ones to the North. The soil should be a fine, gravelly loam. Avoid high, barren soils, and those which are wet. To prepare the soil, make it clean; underdrain, leaving the drain open at both ends to allow the circulation of air. Pulverize at least two feet in depth, making 10 per cent. of the soil, if possible, as fine as superfine flour. For manures, apply 30 bushels of unleached ashes and 12 bushels of lime, slacked with water holding 3 bushels of salt in solution, to the acre. Transplanting should be done with great care, and the rootlets of the plant injured as little as possible. The best time to transplant is in spring, though with care it may be done any time during the summer. The lecturer said he would, in starting a new bed, place the plants three feet apart each way, and allow them to spread till they were only twelve inches from each other.

* Mr. Allen objects to any outline of his lectures on fruit or cattle-breeding being given in this work, as his engagements prevent his revising them.

3

Nearer than this they should never grow. The beds should be mulched with tan-bark, straw, or some such material, to the depth of half an inch—no more. This keeps down weeds, and keeps all but the strongest runners from taking root. Water may be added with great advantage in large quantities, except during the flowering and ripening periods, provided always it does not stand and become stagnant on the soil. After this preparation little attention is needed. The hoe should never be used about the plants, as it injures the roots. Field culture differs little from garden culture. The productiveness of the strawberry about New York does not average more than 40 bushels to the acre. There is no difficulty in raising 150 bushels under the cultivation he recommended. In the winter the plants should be lightly covered.

The strawberry may be made ever-bearing by entirely preventing the growing of runners. This may be done by planting in soil composed of three-quarters river sand and one-quarter woods-mold. This dwarfs the plant and makes it ever-bearing. The staminate and pistillate plants need not be grown within thirty or forty feet of each other. Seedlings are easily raised. The analysis of the plant differs in different places. The best six varieties are Wilson's Seedling, Hooker's Seedling, Longworth's Prolific, Hovey's Seedling, Burr's New Pine, and McAvoy's Superior. There are many others nearly as good. Wilson's Seedling is very prolific; 260 berries, many of them large ones, have been grown on a single plant.

SEVENTH DAY.—Feb. 8, 1860.

When the good Dr. Grant mounted the rostrum yesterday, he was greeted with loud applause; and well he might be, for he has not only the thorough acquaintance with the vine which long years of practice impart, but he bears upon his benevolent face that stamp of integrity which begets confidence and respect. I fear that the audience were but illy impressed with his real knowledge, however, for present sickness has almost deprived him of voice, and the lecture must have been unsatisfactory, because imperfectly heard.

In preface, he alluded to the wonderful growth of wild vines in wet and poor soils, but showed that not only was excessive growth of wood a poor recommendation to the vineyardist, but the quality of wild grapes is poor, and their apparent great yield deceptive. All of the European vines are believed to have sprung from one species, and been introduced from Asia; while in America, the wild vines of the several districts, although widely dissimilar, have not been positively proved distinct species. True, the Scuppernong, with its family of Muscadines, is so peculiar that from its foliage it would scarcely be regarded as a grape. The family of which the Herbemont is a type, is quite distinct from all others, but he believes it to be traceable to a European origin. Many of our native vines have been cultivated with care in the vineyard, but they have not thriven under the treatment so as to recommend them above, or as equal to, the nobler sorts. In vine culture as in other things, the greatest skill and care gives most favorable results. Not a quarter century will pass before the Connecticut farmers, at least those of the southern part of the State, will hail the grape harvest as the most joyous part of the year. Wine-making is an art in which the most complete success can only be attained through much accurate observation, and with great pains-taking and skill; but grape-growing for table fruit is so simple an affair as to be within the reach of any one who will give it the slightest attention. If any one thing in vine culture is more important than another, it is good pruning. Shoots are the growth of one year, and are so called from the time that the opening bud in spring has developed its first leaves, until it has completed its year's growth, and is ready for the pruning-knife. When cut back to one bud, the stump is called a short spur; when cut to three or four, a long spur; and when left with more than this number of buds, it is a cane, except when peculiar circumstances give it a special name. When two shoots spring from a stump near the ground, and are destined to have bearing shoots grown from them, they are termed thighs; and such when laid horizontally are sometimes called arms. The objects of pruning are: 1st. To restrain the roots and branches within convenient limits for cultivation. 2d. To concentrate the strength of the vine, and not suffer the production of useless wood and foliage. 3d. To get just enough wood to bear full crops of good fruit, and plan its distribution with reference to the health of the vine. There are three kinds of buds—the primaries, which come at the axils of the leaves, or where the footstalk joins the shoot, and which in bearing-vines are the fruit-buds one season, and the next produce the shoots on which fruit is borne; the secondaries, which come on the side shoots, or laterals, and which are removed in summer pruning; and the adventitious buds, which are unseen, until they burst through the bark of the former year's wood. They are called wood shoots, as they produce no fruit except in a few varieties of remarkable productiveness. A bunch is a productive tendril; a tendril an abortive bunch. The points or ends of bunches should be cut off, as this causes a complete ripening and sweetening of the upper grapes, and prevents the growing of shrivelled berries at the point, which is a sheer waste of substance. If a vine is left to itself to grow, the tendency of vitality is upward, the fruit gets beyond our reach, has a coarse quality and a woody flavor, while the buds near the ground soon perish, and no after care can revitalize them. It is scarcely possible to fix the duration of a well-set vineyard; it may as well last one thousand as one hundred, or a score of years. The vine needs moisture ever, wetness never. Nitrogenous manures are good if well rotted and composted, for they attract moisture, and a well-prepared grape border is never dry in even the hottest seasons.

In the evening the Doctor was put upon the stand and subjected to a cross-examination of the most rigid nature. Some of Mr. Allen's questions created much good-natured merriment, for he was evidently determined to make our Iona friend, and his sympathizers, give their reasons for the faith which was within them. The information elicited in reply to questions was: That table grapes of first quality could be grown more abundantly and surely 1½° above New York city, than elsewhere in the country. They will not reach so perfect a maturity, perhaps, as in some warmer sections, but they keep better throughout winter, which is of all the most important point. If ripened too early grapes lose flavor, and if the grape-grower is so far north that he is forced to lay down his vines through the winter, he is amply repaid for his trouble in increased flavor and quality of product. The best of the wine-growing region in Germany is that where laying down in winter is requisite. A favorable exposure makes a difference of almost, if not quite, one degree of latitude. The best methods of laying down vary; a mere covering of boards is enough to guard against slight frosts, but with the additional precaution of covering with sand one is perfectly safe in the worst places. But a slight covering is necessary—just enough to guard against having the sand wash or blow off and expose the vine, and two or three inches of depth is enough. The whole vine should be covered. If the vine is as large as a man's arm, it will still readily lie down, if it has been so treated from the first. Milo carried the bull because he commenced carrying it when a calf, and continued the practice. A large vine is not so liable to destruction by frost as a small one. At six cents per pound, an acre of grapes, prepared in the best manner, will yield annually $400, at an expense of $100. For vineyard culture, we can have only 75 per cent. of perpendicular vine area to 100 of surface area of the ground. That is to say, if our vines are set 6 feet apart, they must not be suffered to grow more than 4½ feet high. Sunshine is more necessary to a vine than actual surface-room; and if the vines grow more than the 75 per cent. high, portions will be shaded by the adjacent vines, and thus the crop be damaged. It is a bad plan to bury the bodies of dead animals near grape-vines; they should be composted with three times their bulk of muck, or like earth, the year previous to application to the vineyard. Trenching is good in warm latitudes, because it gives the vine roots a cool, even temperature. Roots should be free to run downward, for if near the surface they get baked to death. In Madeira, vines have an average depth of 7 feet of soil, and grow only on hills.

At this point, Lewis F. Allen spoke of the wonderful growth, hardiness, and productiveness of the wild vines of the woods, and wanted to know why these new sorts, which need so much care and outlay, were their superiors. A gentleman present suggested to him, that if he (Mr. Allen) was content with the quality of fox-grapes and their wine, was willing to go to the woods and climb sky-high to get them, the better sorts were not better for his purposes. But, as the world is foolish enough to prefer the Chasselas, Hamburgh, Catawba, Delaware, and such grapes, to the wild variety, and would pay for a bottle of Hockheimer, Clos Vougeôt, or Johannisberg, more than would buy an ocean of currant or fox-grape wine, these better grapes were better for the cultivator. If we want these splendid wines, we must raise the grapes from which they are made; and, to do this, we must select better soil, give more labor and care to cultivation, and spend more money.

Dr. Grant said, that although thorough drainage was necessary where the soil was naturally wet, yet, if possible, such soil should be avoided for one naturally drained—say a clay loam or a gravel subsoil. Drains, in moderately wet soil, would be likely to get choked with grape roots; but if water were constantly running through the drains, the roots would probably die by immersion in it. He thought that by laying the drain-tiles in, and covering and surrounding them with very poor soil or sand, the grape roots would not pass through it to the drains. The skin of American grapes parts readily from the flesh, and hence in a good table grape may be somewhat thicker than is admissible in Europe, where this free parting is not found. The flesh should be sweet to the very centre, and the seeds should be very small. For family use, where 25 feet length of a wall can be had, the French "Thomery" system is best, but for gardens the simple low "thigh" is perfectly suitable. As it is impossible to fairly describe these systems without the aid of cuts, I refer inquirers to Dr. C. W. Grant, Iona Island, near Peekskill, N. Y.

At 2½ o'clock this afternoon Mr. PARDEE continued his lectures on the small fruits. The raspberry was spoken of first. Few persons, he said, had ever seen a first-rate one. The gardeners about our cities do not succeed in growing them to perfection. This fruit likes a moist, cool situation, such as the north slope of a hill, or the north side of a fence. The soil should be made very rich; you cannot overfeed the raspberry. The strawberry has a multitude of fine fibrous roots, and as it grows little woody fibre it requires little manure; the raspberry, on the contrary, produces considerable wood, and as it has few fibrous roots with which to take up nourishment, these should be well supplied. The soil should be made very fine. Plant about four feet apart, and cut the canes to within one foot of the ground. At the time of planting, stake with strong stakes. Those which will last forever may be made by the French "Burnetizing" process, which is as follows: soak the stakes six or seven days in a solution of blue vitriol and water, in the proportion of one pound of vitriol to twenty quarts of water. Berries raised on canes which have been carefully tied to stakes are much finer than those which have been left to be blown about by the wind. As soon as the raspberries have all ripened, remove the wood on which they grew and allow the sap to flow into the new canes, which will bear another year. Keep the ground clean. In the winter lay the shoots on the ground, and cover lightly with earth. Brincklé's Orange Seedling is one of the very best varieties, and is wonderfully productive.

The Fastolff, Franconia, and Red Antwerp are very fine. Most of those sold as Red and Yellow Antwerps are spurious. The best everbearing varieties are the Ohio Black Everbearing, the Merveille de Four Seasons, and the Belle de Fontenay.

The blackberry may have the same cultivation as the raspberry, and it may also be shaded by trees without injury. Capt. Beverly, of Needham, Mass., introduced the improved high-bush blackberry. The proper way to gather Lawton or New Rochelle berries for the family is, to jar the canes with a hammer, and catch the berries which fall. The others—and these are those sent to market—are not fit to eat. Never leave more than three canes in a hill, and have no suckers growing near the bush, if you want fruit. If you wish plants for sale, do otherwise, of course. Cut back your canes as soon as they have borne their crop, pinch off the ends of the shoots in September, and again in spring; by which plan you will throw the strength of the vine into fruit-bearing on the laterals.

The cranberry, on bog lands to which a dressing of sand has been added, should give fifty bushels per acre the first year after planting, one hundred and fifty bushels the next, and so on up to four hundred and fifty bushels, as a maximum.

The gooseberry is a fine fruit for family use. With me, said Mr. Pardee, it has never mildewed. I know not why, unless it is because I grow them in the tree form, give them clean culture, and in the spring give them abundance of soapsuds.

The whortleberry is difficult to transplant, but with care it may be made to produce abundantly.

The currant is one of our very best small fruits. Like the raspberry, it cannot be manured too highly. Those who cultivate only the Red or White Dutch Currant, do not know what a good currant is. The best kinds grow to the diameter of five-eighths of an inch, and are as much finer in flavor as superior in size. The following are, in my opinion, the best varieties: La Versailles, La Hâtive, Cherry, White Gonduoin, and White Provence.

EIGHTH DAY.—FEB. 9, 1860.

Surely no one is better able to give a valuable lecture upon nursery management than the owner of the largest nursery in the world—no one more capable of discoursing upon horticulture than the ex-editor of *The Horticulturist.* What wonder, then, if Mr. P. BARRY'S lecture this afternoon should have drawn a large audience, and given satisfaction. It is this feature, I think, that gives Professor Porter's Yale discourses great value, that his talkers are workers, his expounders of theory eminent in practical experience. To have Fitch on Insects, Barry on Nurseries, Johnson on Chemistry, and Grant on Grapes, is like having Mott on Surgery, Palmer on Sculpture, Church on Painting, and Greeley on Journalism. And until you can convince me that Paul Potter's bull is of more importance to the nation than Samuel Thorne's Grand Duke, Wedgewood's pottery than the rougher sort which old Mr. Johnston buries underground, I must think that our agricultural lights shine with more useful brilliancy than would those at the supposed convention of savans and artists.

Mr. Barry commenced by saying that, although the subject of nursery management might be deemed not generally interesting, since it was a calling by itself, yet every one who intended rearing an orchard, or even a few trees upon his farm, should know enough of the mode of managing trees to rear what few he might need to supply deficiencies which might arise from death or other accidental causes, or at any rate to give to his growing orchard or plantation such good care as would make it most profitable. Twenty years ago, two or three small nurseries in the neighborhood of each of our large cities, occupying in all not more than five hundred acres, and a few other small apple nurseries of an acre perhaps each, supplied the wants of the United States and the Canadas. Now we have over one thousand nurseries; and in Monroe county, N. Y., alone, where he resides, there are three or four thousand acres, producing annually $500,000 worth of trees. In the whole Union, there are annually sold fifteen to twenty millions trees, for say $5,000,000. His subject he would treat under the several heads of *locality; soil; arrangement; preparation of the ground; propagation of stocks; grafting; treatment of trees in the nursery;* and *digging up.* A commercial nursery should be located near a large city, town, or village, both for the facility of getting a supply of labor, manure in abundance, implements, post-office, and railroad, or other transportation ; and a preference should always be given to a fertile and prosperous agricultural region, for obvious reasons.

Surface.—The surface of a nursery-ground should be nearly level; if sloping, the slope gentle and nearly uniform, not only for the convenience of working and planting in straight lines, but because hilly ground is so washed in rains as to do great damage. *Shelter.*—There should be, if possible, some natural shelter—high ground, woodland, or orchards, to break the force of winds in winter and spring. If these natural shelters cannot be had, plant parallel belts of rapid-growing trees, such as spruce or larch, in the form of hedge-rows, at a distance of two hundred or three hundred feet apart, all over the grounds. *Soil*—should be dry and deep, neither too light nor heavy. Light sandy soils require heavy and frequent manuring, and produce weak trees; and retentive clays give too little fibrous root to trees, ripen them badly, make transplanting difficult, and good removal almost impossible. Stony soils impede the progress of tools, and are in every way objectionable. On dry soils, naturally drained, trees mature their wood well, and are therefore hardy when transplanted. The coarse-grained, rank, watery trees grown on prairie soil, freeze to the ground in a temperature that would not affect those grown on more favorable ground; it being the fluid, and not the solid parts of a plant, which are acted upon by frosts. A nursery needs much more thorough drainage than ordinary farm fields. The drains should be never more than two rods apart, and were better to be laid at a depth of three and a half feet. In a stiff, retentive clay bottom, they should be only twenty feet apart. *Laying out the Nursery.*—Divide and subdivide your land into plots and compartments for the various articles which are to be grown; assigning special places to seedlings, stocks to be worked, cuttings, layers, and specimen trees. This latter plot is very essential to the proper management of the nursery, and the comfort of visitors. In this specimen plot should be grown one or two samples of every tree in cultivation in the nursery, the better to test their genuineness, quality, and constitution. A place should also be given to manures and composts; and through the whole nursery broad roads should be made so as to make every part accessible. *Preparation of Ground.*—An old pasture, or clover field, is best for nursery ground, for the inverted sod gives just the right food for young trees. A broadcast, light dressing of well-rotted manure, or compost, should be applied before plowing. Plow very deeply, and subsoil fifteen or eighteen inches, if possible. This roots your trees well, lets surface water run down, and lower moisture draw up, and in fact is every way requisite. *Propagation.*—Our cultivated varieties of trees cannot be propagated by seeds. The particular qualities which constitute their chief value are the result of hybridization, or of cultivation—qualities which are not transmissible in the seed. True, we may chance upon better varieties by sowing the seed, but there are a thousand chances against such good fortune; and hence we resort to grafting, budding, cuttings, layers, and suckers. And this brings us to the subject of stocks, which is a most important one in the propagation of fruit-trees.

Without good stocks we cannot produce good trees, although our soil, situation and cultivation may have been ever so favorable. Formerly, wild, self-sown seedlings from the woods and orchards were thought good enough for the nurseryman's purposes, and even poor suckers from the roots of trees were used. Experience has taught us better practice than this, and now the production of good stocks is the first great aim of intelligent cultivators. The apple, pear, plum, cherry, peach, apricot, and nectarine stock, are grown from seed; but the *Doucin* and *Paradise* for Dwarf Apple-trees, and the *Quince* for Dwarf Pears are usually produced from layers. We have thus far been able to grow cherry and common apple stock in sufficient quantity for our use, but are compelled to import pear and plum seedlings and stocks for the dwarf pear, apple and cherry. The most important of all these is the pear, which we have to import largely, because in this country the young seedling is attacked by a fungus or blight which destroys it at a tender age. Although no absolute remedy for this "leaf blight" is likely to be hit upon, very thin sowing of seed on a deep, dry, fresh soil never before occupied by trees, and unremitting care and good cultivation during the early stages of growth, act in some wise as preventives against the malady. Our nurserymen now grow on one acre as many seedlings, especially the apple, pear, and plum, as should rightfully be assigned to five, and the result is, a growth of weak, spindling trees. Well-grown pear and apple stocks should be always ready for the nursery rows at one year old. If they are not, another year's occupancy of the same place will not generally add much to their value. Apple stocks may, perhaps, remain two years in a place, but pears must be transplanted. The lecturer then described the stocks in common use for grafting, dwelling for a moment to sketch the difficulties which attended the introduction of the quince stock for dwarf pears into this country. Experience has established the fact that the two French quinces, the Angers, and Paris or Fontenay, are best for pear grafting. The former is most vigorous, and of rapid growth when young; the latter more hardy. Some pears succeed best on one, some on the other. Stocks are good when half to three-quarters of an inch in diameter, and can be obtained from cuttings, layers, or by the earthing-up practice. To obtain strong stocks, plant out a certain number of stool or mother plants, in a deep, rich, well-prepared soil; when they have stood one season, cut them all off close to the ground. The next season they will produce strong, smooth shoots, which the following year may be earthed up, half their length, as celery is earthed up, and in the fall they will have rooted well enough to bear separation from the parent plant. If left on during winter, the frost will ruin them. Such stock as these may be set in nursery row the next spring, and budded the same season. Only two crops of shoots can be taken from the same stool, and a good dressing of manure is necessary to get even the second. Pears propagated on small, weak quince stocks are worthless. In budding or grafting quince stocks, it should always be done near the ground, so that the whole of the quince may be set under ground without being too deep. Root-grafting, although still an open question among nurserymen, Mr. Barry believes to be, if properly performed, as good a mode for propagating the apple, and more especially all the strong growing sorts, as any other in use. It has been sadly abused, and thus been brought into disfavor with bunglers and their victims.

Management of Young Trees.—Trees are too closely planted, as a general thing; three and a half feet between the rows, and three or four inches between the plants, is too little space to give either air, light, hardiness of constitution, spread of root, or strength of top. For apples, pears, or other trees which are to remain two years in the nursery row, the distance from tree to tree should never be less than eighteen inches for standards, and twenty-four inches for pyramids; and even at such distance the pruning-knife is to be freely used. Country people are too apt to value a nursery tree in proportion to its height, rather than its strength and proportions—a too common and fatal mistake. Cutting back should be freely practised, and the leader or main stem should be pruned as well as the side branches, else one will get a tall and ill-proportioned tree. An enormous amount of money is annually lost to tree purchasers from rude and unskilful taking up. Trees are torn up by the roots, as if the trunk and branches were the one thing necessary, and the roots superfluous. The proper way is, to open a trench on each side of the tree with a common spade, *keeping the edge toward the tree*, so as not to cross a root. These trenches should be far enough from the tree to avoid the main roots, and deep enough to go below all, except a tap-root, which may be cut off. This being done, the tree may be pulled up with its roots entire.

Mr. Barry, in conclusion, spoke of the wide field which was still open to intelligent, industrious, and capable men, who would embark in the nursery business, but cautioned them against entering upon it for mere speculative purposes, or with dreams of sudden wealth, to be got as one would draw a lucky number in a lottery.

The morning lecture was by Prof. Johnson, and the one after Mr. Barry's was to have been by Dr. Grant, but as he was too much indisposed to speak, he procured as a substitute Mr. ANDREW S. FULLER, the Brooklyn nurseryman. Mr. Fuller went into the history of the grape in Europe, noticing the varieties which in successive ages were deemed the best. He showed when and how these foreign varieties were introduced into the United States. In the Northern States they had, almost without exception, proved failures, but at the South they had given rise to descendants, some of good quality. Even with a choice grape, its quality and profit depended in a great degree upon the cultivation and pruning given to it. In summer, during the season of active growth, the liquid portions of the sap are exhaled almost as fast as they can be absorbed by the roots, and no great accumulation can take place in any one portion of the vine. But the leaves once fallen, the roots continue to absorb their appropriate food from the soil, and thus the wood becomes quite filled with sap, which is kept in store for early spring use. It is therefore plain, that we should prune our vine as soon as the leaves drop of, that the sap which is afterwards absorbed may all go toward the nutriment of the buds which remain.

He recommends a medium depth of planting; that the surface or upper roots may be not less than four nor more than eight inches from the surface of the ground. Many of our strong-growing sorts, such as the Concord and Diana, can be brought within control by root-pruning for the two or three years after planting. Mr. Fuller thought that if we may judge from our short experience, we are warranted in the belief that America will produce, if it has not already, as fine grapes for both table use and wine-making, as the most favored countries of Europe, with all their centuries of experience, can boast.

NINTH DAY.—Feb. 10, 1860.

Whenever, in coming out of a lecture-room, you hear all about you people saying "What a capital lecture!" "How well he understands his subject!" "How many valuable hints he gave us in the hour!" you may be certain that it was a valuable discourse; and such was the case this morning, after Mr. Barry's second lecture on fruit-trees. Certainly I never listened to a more complete epitome of information on any one topic than he condensed into sixty-five minutes; and now that I sit down to give your readers the gist of it, my trouble is to know where to commence the process of exclusion.

The subject chosen was the "Transplanting and Management of Trees in the Orchard and Garden," embracing a variety of operations which, if followed in detail, would require a week instead of an hour to describe. The general remarks upon the preparation of ground for nursery trees, which were contained in my letter of yesterday, apply to all tree plantations. Our readers should remember that the important points in land treatment can only be best done before the trees are set out; so that before we send our orders to the nurseryman we should have finished our draining, subsoiling, and trenching. As to spring or fall planting, opinions vary, and vary chiefly because of different nature and conditions of soil with various tree-planters. Mr. Barry's experience is, that in a good, dry, well-prepared soil, fruit-trees may be planted at any time after the wood is ripe in the fall (a period indicated not by the fall of the leaf, but by the perfect formation of the terminal leaf-buds, and the changing tints of the foliage), until the freezing of the ground; and, in spring, from the time when the frost is out and the ground dry enough to work, until the buds have made some considerable advancement toward opening. Generally the more tender trees, such as the peach, apricot, and nectarine, should at the North be planted in spring, as winter acts severely upon them after transplanting. This is the better mode, but fall planting of even these tender, juicy-wooded trees, is often successful, if precaution be used. The fall planter must never forget to mulch the roots with several inches' depth of leaf-mold, half-rotted manure, or some such material as will modify the action of frost on the roots and tree-trunk. A neglect of proper preparations for planting causes great loss. The majority of trees from the nursery, by unskilful removal, have mutilated roots; if the tree were set without proper pruning, most of these roots would rot, and those which escaped would grow feebly for a long time. All these bruised and broken roots must be pruned close up to the sound wood with a sharp knife, the cut being made perfectly smooth and almost straight across, so as to present as little surface as possible. Never cut the roots downward, or so as to have the slope on the upper side of the wood, but upward; for in any other case the water would get between the bark and wood and rot off the root, while if rightly done new rootlets will be put forth from the root end, and all go on well. All broken branches must be removed, and then the whole top be reduced by cutting back half, or more than half, but always keeping the lower branches of dwarf pears and other pyramidal trees, longer and stronger than the upper ones. The tree naturally pushes its growth upward, and this tendency must be restrained so that you will get the bulk of fruit near the ground, thus avoiding top-heaviness, and liability to branch-breaking by high winds. Keep a due proportion between root and branches, so that there will always be enough root to furnish food, and no waste of strength in superfluous wood and leaf-production. We aim at getting fruit in large quantity, and of distributing it equally over the tree, that no one part may be overtaxed, or weakened. Almost ninety of every hundred tree purchasers set such store by the nice, long, smooth branches of their trees, as they come from the nursery, that they spare the knife, and set them out just as received. Let them beware how they are thus "penny wise and pound foolish," for their trees are checked and stinted in growth, and are left far behind others which have been boldly and judiciously pruned. Many persons think trees should be manured, like a hill of potatoes, at time of planting. Such are likely to kill their trees by overmuch kindness. Good fresh surface-soil—if light and sandy, all the better —is what should be put around trees at time of planting. He would say nothing about hole-digging, for the whole soil where trees were to be planted should be so well prepared that a hole needs only be large enough to admit the roots. The roots should be set about four or five inches below the surface. In light soils they may be set deeper than in heavy ones, because heat more readily passes downward. The thorough cultivation of the soil among fruit-trees can be neglected only at the planter's peril.

In fields of grain the poor trees are smothered by their avaricious, or unwise, owners. When the rows are thirty or forty feet apart, almost any farm crop may be grown between, but at least six feet of ground beyond the extremities of the roots should be unplanted, and kept as clean and as mellow as it would be about a hill of potatoes or corn. No weeds must exercise Mr. Douglas's squatter sovereignty privilege, unless one wishes to starve his trees to the extent of the food these pestiferous plants consume. Remember this point, for it is of the utmost importance; but in putting it into practice, remember also, that in your hand-hoeing, or horse-hoeing, the tree roots must not be disturbed. A light annual dressing of compost should be spread upon the surface early in winter, and in spring forked in. Road-scrapings, ditch-bottoms, and such matter, are good for application to a light soil, and heavy leaf-mold, and decaying vegetables, with stable-manure for a heavy soil, are good in compost. Occasional light dressings of lime, ashes, and even salt, will be found beneficial. Mulching in summer should be very light, just enough to keep down weeds, and once a week, or once a month, as the case may be, must be removed for as thorough a forking of the ground as can be given without injury to the tree roots. The object sought in pruning fruit-trees is to regulate their growth and bearing, so as to secure at once a particular form with greatest vigor and fruitfulness. The only instrument used in a good nursery is the pruning-knife; and this should be kept so sharp that any ordinary branch may be lopped off at a single draw, leaving a perfectly smooth surface. Shears should never be used. A saw is only required when trees have been neglected. Branches removed should be cut close to the trunk, so that the tree may not be injured by the decay of a stump. Shorten shoots to a good strong bud that will make a leader, not too close to nor too far from the bud, and with a slope of cut of about forty-five degrees. In shortening your leader, don't always cut on the same side, for you would thus make the whole tree lean one way or the other. Pruning, rightly done, is a blessing; wrongly, a curse.

To show practically how pruning should be done, Mr. Barry performed the operation on several fruit-trees which he had brought for the purpose, and I have no doubt but that the large audience got thus a far better idea of the *modus operan-*

di than long arguments would have conveyed. I am also glad to learn that Mr. T. S. Gold intends to illustrate his lectures on sheep-breeding, by placing before us a well-shaped and a badly-shaped live sheep. Could anything be more admirable?

Standard apple-trees in orchards require very little pruning. If the head is formed at a proper distance from the ground, say four or five feet, and the main branches to form the framework of the head are started in the right direction, as nearly as possible equally distant, inclining upward and outward, the subsequent pruning will consist in removing branches where they are likely to become crowded or to cross each other. The natural growth of varieties differing, our pruning should be modified to suit each special case. Apple-trees not pruned generally bear a heavy crop of fruit one season, and none the next, and so heavy is the crop that a good part of it is worthless. Judicious pruning enables us to have a moderate crop of fine fruit each year, besides promoting the general health and prosperity of the trees. A few days of a man over an apple orchard when the fruit is half or a third grown, will be well spent in removing misshapen and wormy fruit, and thinning out clusters that are crowded together. Fools cut away branches indiscriminately, until their trees are but skeletons, with a few bearing branches at the extremities only. The force of the tree is then expended in producing a crop of rank, watery shoots in the interior, to be again cut away to make room for a second crop. Trees should never be suffered to bear fruit until they have got strength and vigor. A pruner should know the difference between fruit-buds and wood-buds, and at least the rough outlines of the principles of tree growth. This knowledge may be acquired by an intelligent man in a brief time. There are many other points of equal interest in Mr. Barry's lecture of which I should like to speak, but cannot.

Doctor Grant lectured first this afternoon, speaking without notes, and, like Mr. Barry, exemplifying the doctrines of pruning and vine-setting, on specimens brought for the purpose.

The following directions he gave us for preparing a grape-border in the best manner—(our readers will remember that the term "border" is applied to any plot of ground longer than wide, which is to be devoted to grapes):

For a trellis of vines, more than twelve feet of width is unnecessary, and one-third less will answer very well; and it is desirable, but not indispensable, that half of the twelve feet should be prepared before planting. If only a width of three feet is prepared, three feet more should be added the next season. To prepare the border immediately, the unfertile soil that lies beneath must be removed, and fertile soil put in its place. To do this, a trench two feet wide is made to the depth of the mold, or fertile soil, which we will suppose to be one foot; if more than that, so much the better. Now, to make the border two feet deep, which is the least admissible, one foot of the subsoil must be removed. If grounds are of considerable size, this may be spread over the surface of a portion, so that it shall not be more than two inches in depth, and plowed or worked in without any immediate damage, but with ultimate benefit, particularly if manure is used at the same time. Into the bottom of this trench the fertile soil of the adjoining two feet is put, and, if it can readily be had, a compost of leaf-mold, or muck, or any vegetable decay, and well-rotted stable manure, thoroughly mixing the mass as it goes in. If sods from a rich pasture can be had, they may be thrown in with the compost to the depth of fourteen or sixteen inches for every foot of subsoil removed, and then the fertile soil from the next two feet put upon the top. Repeat this process until the border of required dimensions is made, and finish by putting into the last trench the soil that was taken from the first. If sods and compost are not used, other fertile soil must be obtained from adjoining ground, or some other quarter, to replace the subsoil that has been removed. At the completion of the operation, the ground of the border will be found to be some inches higher than the adjoining ground, but in two years it will settle to the level. This is the operation called trenching, and without it no garden is in condition for giving best results. For growing strawberries, raspberries, and blackberries, it is equally advantageous, but with this difference, that the fruits last named are expected to continue perhaps only from six to twice six years on the same ground, while vines properly planted and managed have no limit to their duration, and the fruit for many years will constantly improve in quality and earliness of maturity. If the trenching is performed one season in advance, the subsoil may be put upon the top of the mold, and enriched by having manure thoroughly incorporated by a second or third spading, or by plowing, according to extent of ground. If ground is prepared in early autumn, it will be ready for vines in the spring; but if in spring, it will not be in the best condition for vines before fall, without a renewal of subsoil.

The subjects of pruning and planting were also fully discussed, but my space is already exhausted, and I must leave them undescribed.

In the evening, Mr. George B. Emerson, of Boston, gave a lecture upon "The character of the various forest trees of Europe and America." He alluded, in commencing, to the differences observed in the tree of the plain and the forest: the one tall and bare, the other full of limbs, and short. He then went on to speak of the great uses of the forest in creating soils. Described the lava-covered sides of Vesuvius, where the lichen first, the moss, the grass, the low shrub, small trees, and finally larger ones, added to and made the soil upon which grows the tree of 400 or 500 years. One office of the forest is thus to prepare a soil for the use of man. As forests have disappeared here, we have an unfavorable change of climate, becoming colder in winter and hotter in summer, and the streams become dried up. Many places, in valleys once protected, are now open to the cold blasts, and nothing will grow well. A row of trees planted across the valley would mitigate the result in one generation. He considered the more extensive means to protect by means of trees, in France, Germany, and England; alluded to the advantages of forests as electrical conductors and condensers thus of moisture; spoke of the vast stores of the sun's light and heat they annually store up, to light up the long evenings for man, and of the denudation or washing away of the soil when the roots of the sturdy trees were gone. —Restore the forests to the tops of our hills, and the moisture would be restored to our air and droughts prevented.

TENTH DAY.—Feb. 11, 1860.

The convention assembled at 9 o'clock, and listened to another lecture by Mr. Emerson. The number of ladies in attendance was larger than at any previous session. The subject was "The Individual Trees of the Forest." In introducing it, he remarked that the feeling was common that the farmer's was not a high occupation. There is no occupation requiring such large resources of knowledge. Man can only prepare himself for the proper culture of forest trees by studying them in their native woods. The cultivator of the forest tree must have varied knowledge;—of physics, in their higher departments, treating of climate—for we can do a vast deal to change it most favorably; of the sun's light and heat and their action,—a lesson seldom learned as it should be; of electricity and the kindred forces; of the winds and the waters; of the chemistry of soils and the proper action of their elements; of oxygen and hydrogen in water and the other organic elements found in trees; of the laws of the atmosphere, in winds, rain, and dew; of the operation of manures and their adaptations. The forester must know what soils will furnish the necessary nutriment, and to this end must know the composition of trees. Structural botany, one of the most curious sciences that the genius of man has laid open, must be understood. So of endosmose and exosmose, the strange forces by which food is taken into the plant; of the composition of the products of plants, the formation of wood, and the circumstances favorable to growth; how to manage the ground in preparation for planting, to select the place where they shall flourish, and the trees of the best form for planting. He must be acquainted with the friends and foes of each tree, both insects and birds, and with the various processes of layering, budding, grafting, &c. The observant faculties are all necessary, and ought to be educated; their neglect is among the most serious omissions in a farmer's education The objects of forest culture are to improve the land and to furnish materials for use in the arts. Of single trees, those are best which will furnish the shade we seek. The spray of some of our native trees, as the birches, and willows, and especially the maples, is most beautiful, and varies every season of the year, ever being a source of beauty. The seeds of different trees fall at different times, according to their size, so that they may be covered up and germinate, and generally under the shade of the mother tree. Seedling trees must be sheltered, a purpose for which the Scotch or other fir is used to good advantage. The value of leaf mould for these seedlings is well known. The ground for the seminary, or nursery, must be well prepared, but need not be very deep. The best manures are leaves and leaf-mold, with a little of barn-yard compost, well rotted, and then all suffered to lie for a year exposed to the air. The seeds ought to be sown immediately upon gathering, those which animals would dig up for their food excepted. Many seeds will not bear drying. In imported seeds, some few come up the first year, some the second, and others the third, whereas had they been immediately sown as they were taken from the trees, they would all have sprung up at first. The depth of seed planting varies according to size, and the young trees must be protected for the first year or two, from the sun. By transplanting we cut off the tap-root, and thus render it easy to remove again. Each soil as we advance should be poorer, till it becomes of the same character as that where the tree shall stand. Trees for an artificial forest should grow close together, and single ones apart from each other. The oak, which has been allowed to expand, is one of the most magnificent things on the face of the earth. It is a singular fact, that some of our finest trees are not to be seen growing in our own forests in their native perfection. To see our scarlet oak in beauty, we must see it on an English lawn. The nurseries should be kept free of weeds by the hoe or rake. No small part of the success with trees depends upon the care with which they are taken up, and also upon the shortness of time they are out of the ground. The rootlets are killed often, if dried by the sun or wind, and the tree has to throw them out anew. They may be planted on the lawn in rows, singly, or in groups. The land should be trenched, and supplied often with bones and ashes, the trees needing both phosphoric acid and potash. A singular fact made known by the united researches of chemistry and microscopy is, that only in a liquid containing sugar, dextrine, and protein, can cells be formed. Only where carbon, hydrogen, oxygen, nitrogen, and often sulphur and phosphorus exist, can the first act of plant life begin. Plants generally contain three per cent. of nitrogen. This must be added if the soil does not contain it. Mulching with leaves, sedge, grass, or rotten wood, is advantageous. Is it asked, What trees are best for the lawn, or near a dwelling-house, for the pasture, the public square, or the road-side? Every tree is more or less beautiful. Every tree is a picture, varying in color, shape, and all the accidents of vegetable life, in all the hours from the beginning to the end of the year. It may become an heir-loom, and ever fresh with the memory of parents and grand-parents gone before. Each tree has its birds, and insects, its epiphytes, parasites, and lichens. The grandest tree in our climate is the oak, and the longest lived. In the forests of Massachusetts there are twelve species. The white oak, for the forest and lawn, is susceptible of magnificent development.

The old oaks of the forests and lawns of England are worth a voyage across the Atlantic to see. No language can give an idea of their beauty and grandeur. The English elm is best for narrow ways, the American for broad. The former, though not so graceful a tree, throws out its leaves earlier and holds them later, being in foliage from three to six weeks longer. The elm can speak for itself, for it is the only tree that everybody knows. The tulip-tree, a rapid grower, with fine flowers and fruit; the sycamore; the Norway maple, standing the wind better than any other tree; the red, white, and rock maples, the last the best; the beech, with its showy blossoms and sweet nuts, good for pasture, because never struck by lightning nor browsed upon by cattle; the linden, and hickory, easy to transplant if the tendency to depend on the tap-root be corrected in the nursery; the sassafras, hornbeam, hop hornbeam, the locust, the horse-chestnut, and black-walnut, all have their advantages. Two or three black cherry-trees along the outside of a cherry orchard, will draw the insects to themselves. The plane, or buttonwood, makes a conspicuous figure in all grounds, and was valued by the Greeks and Romans above all other trees. Birches are admirable, too, for the beauty of their bark, leaves, and branches.

Professor Johnson gave us a capital lecture on the nutrition of animals. The food of man in his best development, says the Professor, is not exclusively vegetable; not but that from vegetables he could get all the substances which he needs for his sustenance, but in the form of flesh they are much more condensed.

The animals which exhibit the most intense power of muscular and nervous force are carnivorous. For the sake of flesh and milk as food, for wool as clothing, and for the useful labor which the ox and horse furnish, the farmer seeks to convert vegetable into animal produce. By the aid of cattle, not only can man convert the grains, fruits, and esculent roots into a more concentrated and vigorous diet, but he can manufacture food out of naturally growing grasses, and employ hundreds of otherwise refuse matters for the same object. A diagram exhibited by the lecturer showed the composition of a pig when fat and lean, thus:

	Fat—Per Cent.	*Lean*—Per Cent.
Water	45	60
Albuminoids	15	17
Fat	37	21
Mineral matter, or ash	3	2
Total	100	100

The carbo-hydrates—starch, sugar, cellulose, gum, &c,—are changed by the animal into grape sugar, and are then ready to be assimilated to build up its body. The grape sugar is changed into lactic and butyric acids, and thence into fat. The mineral matters found in the bones, blood, and other portions of the body, are of course obtained from the plants, which in their turn suck them from the soil. In some districts, such as that about Leipsic, some of these necessary minerals are deficient in the soil; and it has often been observed, that where phosphate of lime is not in the farm soil in sufficient quantity, cows suffer from bone disease, and will gnaw any old bone that they may find lying on the ground. Animal force and heat, like steam, are generated by the actual combustion of material; in the former cases this being food, in the latter fuel. The "fire-place" in the animal is all over its body, wherever a pin-prick will draw blood. As in the steam engine, the amount of muscular and nervous force in the animal is proportionate to the amount of fuel or food consumed. First, material is stored up in the tissues for use, and then every exertion of the muscles or brain is accompanied by an oxydation, or burning of the tissues. In this process, carbonic acid, water, and a small quantity of ammonia, are given off—the remainder of the ammonia being transformed into urea, and voided from the body. An engine is merely a mechanism for using an engendered force, but the animal is itself consumed, and must be renewed constantly. Whenever the time arrives that the vital force is not enough to supply the waste, decay, and then death, come upon us. A degree of heat that would destroy animal tissue, when separated from the animal, is necessary in the body to sustain life itself. This heat is engendered by using the carbo-hydrates and fats of food; but these contain no nitrogen, and hence they will not strengthen our bodies, although they do warm them. When in a state of rest, the muscular and nervous tissues are but little wasted, but the fat is consumed in heating. When, however, an ox or man labors, or a man thinks, the muscular and nervous substance is consumed. A good warm stable, or other means of giving external heat to our animals, is a much cheaper way to maintain the requisite animal heat than to overfeed with corn and oats. Oil is a necessary ingredient in food, and the addition of fatty matter, when not naturally present in sufficient quantity in it, helps digestion, and thus promotes the growth of the animal. A German farmer proved this by feeding some stock on food that contained but little oily matter, and comparing their daily weight with the greater weight they afterward attained when fed upon a more fatty diet. For man's food, cooking is a great assistant to digestion, for it commences chemical changes which would have to be brought about in the stomach, and would thus abstract from his store of vital force a considerable amount of what he might have used in muscular exertion. The young growing animal needs an easily digestible food,—food which contains a large amount of bone material. Milk is by analysis found to be of just this character, and hence we see the admirable provision of nature in this respect.

The afternoon lecture was by Mr. Luther H. Tucker, of *The Country Gentleman*, who, having devoted the whole of last summer to an investigation of British and French farming, was deemed the suitable person for giving us a lecture upon this interesting topic. Mr. Tucker is another of our rising young men, and already gives promise of doing much toward bringing about the needed reform in our farm practice.

Mr. Tucker commenced with some remarks upon the English climate and soil. The former is such that while, on the one hand, Indian corn will seldom ripen, and the pear, the peach, the tomato, the melon, and cucumber, and similar fruits, require artificial heat to effect their perfect development; on the other, there is not a month in the year when the plowman and his teams are not actively at work. Of the soil, it had been said that while comparatively little is really very good, one-thirteenth part resists all attempts at cultivation, and two-thirds of the remainder is so stubborn and ungrateful that it tries the skill and ingenuity of the cultivator. He then spoke of the progress which Great Britain has made during the last half century, in population and wealth. A recent report of the Registrar-General showed that the natural increase of the former now averages over one thousand souls every twenty-four hours, while the growth of the latter may be estimated from the computation published a year or two since in London, that the grand aggregate profits of English industry amount each year to *two hundred and fifty millions of dollars* ($250,000,000). There is a national predilection among all classes of the people for country life—a kind of taste which it might be hoped that we should prove to have inherited, when the fever of our younger life should make way for more of the discrimination of cooler manhood—a taste there manifested not only by the attention with which men of wealth regard horticultural embellishments, and the interest taken by Parliament and the whole country in equestrian improvement, including the races, and in sporting—but also in the more practical direction of actually increasing the productive power of the land. So important did this taste appear to Lavergne, the French author, that he did not hesitate to pronounce it "the chief cause of her [England's] agricultural wealth." Prince Albert's farming was referred to as an example in point, as well as the expenditures often made in the cause of agriculture by wealthy gentlemen and commoners.

In a general view of "English agriculture," then, if there were not practical lessons afforded for immediate imitation, a pervading influence could not but be felt throughout, calculated to lead our farmers to a more intelligent appreciation of their calling and its duties. The first cause of its advancement was undoubtedly the abundance of wealth and the compact population of the island. Next came the national taste to turn this wealth into rural channels, and thirdly, a necessity for enlarged production, which had directed both wealth and taste to practical objects. Up to a period within forty years, the object in view by English agriculturists had been to reclaim waste lands. A Committee of the House of Commons, in 1797, after protracted investigations, calculated the area thus brought under inclosure during the eighteenth century at about 4,000,000 acres; and under the impulse of war prices from 1800 to 1820, there are statistics to show that 3,000,000 acres more were added to the dominion of the plow. Then came a falling off;—comparatively little has since been done in this direction, and, since 1840 particularly, the aim of English agriculture has been, not to enlarge the productive average of the island, but to increase its acreable production.

Some of the agencies by which this had been partially, and was constantly being more fully accomplished, he hoped to illustrate before concluding. Previously he alluded briefly, in the fourth place, to the three classes engaged more or less directly in English agriculture—the proprietors, the tenantry, and the laborers. "England and Scotland," wrote Philip Pusey, so long the editor of the Royal Agricultural Society's *Journal*, "are the only countries with a class of cultivators possessing sufficient capital to stock farms of a good size at their own risk, paying a certain yearly sum to the proprietor." In fact, the farming capital, other than the ownership of the land, is almost wholly in the hands of the tenants, and, in many instances, even the park-grazing about the mansion of the landlord is let out, as well as the arable land. The tenants are often men of such wealth that they would probably live upon their resources in this country, except so far as they might be engaged in looking after their investments. The average interest obtained by landlords upon the cash value of their land, high as the rents appear to us, perhaps rarely exceeds three per cent. Farmers expect to invest their money in agriculture so as to make it pay them ten per cent. if possible;—the average profit they realize may vary from eight to ten per cent. Really, it is only a very rich man who can afford to own land in England, and several instances were given to show how property there gravitates toward the country, including a farmer mentioned by Mr. Colman, who was paying an annual rent of $35,000!

Taxes and tithes are to be added to the rents the farmers pay, these rents varying from a dollar or two per acre, under the least favorable circumstances, to ten, twelve, and fifteen dollars for choice locations in good farming districts, and reaching for the whole island an average of six dollars. Some of the Scotch moors are rented according to the number of sheep they will carry per acre, at so much per head for the sheep.

During the eighty years preceding Mr. Caird's investigations in 1850–51, it was found that the rents of twenty-six counties had increased a little more than one hundred per cent., while the wages of laborers showed an advance of thirty-four per cent.; the price of bread was about the same; meat had appreciated seventy per cent., butter one hundred per cent., and wool still more. The production of wheat only showed an advance from the average of twenty-three bushels per acre reported by Arthur Young, to that of twenty-six and a half reported by Mr. Caird—an explanation of which is found in the fact that only the very best fields were then put into wheat, while now the area on which it is grown is immensely increased, and the whole, bad and good, made to yield fifteen per cent. more than the selected parts did previously.

A brief recapitulation of the measures to which English agriculture probably owes its progress was then given,—including among earlier improvements, root crops, rotation, the sowing of grasses and clovers, and the imperfect drainage of the land by open ditches or otherwise, as the most prominent; among later ones, the increased use of machinery and better implements, purchased fertilizers, and food for stock; the deeper drainage of the land by tile and pipe; and, perhaps most prominent of all, the improvement effected in the different races of domestic animals, and the increased attention given to feeding them for the sake of their manure.

A brief account followed of a visit upon a Hertfordshire farm where one of Fowler's steam-plows was in operation. The Norfolk or four-course system was there practised, extended sometimes over a fifth year by retaining the clover-crop a second season, or, if the land was in good order, by adding a grain crop, generally oats. The remainder of the hour was devoted to a narrative of some of Mr. Mechi's modes of farming—an account of his method of feeding, stabling, and managing his manures, and a statement of the crops he has obtained at Tiptree Hall.

Mr. Mechi went on to this farm fifteen or twenty years ago, when he gives the place rather a hard character. "Almost surrounded by barren heath," he found the land so retentive of water that a large part of it was constantly in a state varying in consistency "between putty and bird-lime, according to the season." He sold half of it, determined to get as much as possible out of the remainder, and went on to make such expenditures as really frightened sober and practical men; he underlaid his fields with pipes, conveying manure in a liquid form by means of hydrants to every part of the farm; and now, not only all the stable manure, but also guano is distributed by steam pumps through this channel,—and even the carcasses of dead horses and cattle are put into the same tanks, macerated by degrees, and sprinkled out through the hose.

His average crops of wheat are now forty-six to forty-eight bushels per acre; of oats, not far short of ninety; of barley, not more than fifty. His cattle are fed on sparred floors, without bedding of any kind—all their food cut and cooked. He says that he don't use straw more generally in feeding because it is not naturally in condition to be very nutritive; but when cooked, he states that every hundred pounds of straw is shown to contain the equivalent of eighteen and a half pounds of oil. Straw for manure is worth to him only two dollars thirty-three cents per ton, while for fodder it is worth five dollars; and, as he raises about two tons of straw per acre, this difference is of enough consequence to him almost to turn the scale between loss and profit upon each acre under a grain crop.

Mr. M.'s cooking apparatus consists of "a number of cast-iron pans, or coppers, each capable of containing 250 gallons," set in brick-work, so as to stand level with the floor, and heated by waste steam, from the engine, admitted into a four-inch space about them. The fodder is cut to quarter-inch length, at a cost of from 50 cents a ton for cutting hay by steam to $1 per ton for straw. In feeding roots, they are first cut by machine, and then "mixed in the manger with the warm steamed chaff."

As to rotation of crops, Mr. Mechi, in common with most of the "high farmers" whom the speaker had met, apparently regarded this as altogether a secondary consideration *after a farm once attains a certain pitch of productiveness.*

The difficulty which high-farming is most puzzled to overcome is, the "laying" or lodging of the crop. The moment the condition of the land reaches a certain point, its yield can be no farther increased, because the amount of soluble silica to glaze the straw appears to fail, and the risk from this cause, together with the difficulty of keeping the ground clean, presents an obstacle nearly or quite insurpassable.

An extra lecture was given in the evening, by Mr. Quinby, on Bee-Keeping. In the first place he proceeded to answer the Yankee's characteristic question, "Will it pay?" By a very conclusive array of facts and figures he demonstrated that with intelligent management no investment is more remunerative.

The best season for moving bees is from the first of October to the first of April. They should not be moved during the summer, as the bees will, many of them, leave the hive, and the combs being soft are liable to injury. In preparing for moving, the apparatus should be covered with muslin, and the hive inverted to prevent the combs from becoming detached.

In purchasing, see that the hive contains sufficient honey to carry the bees through the winter, perhaps thirty pounds, and also that you have a large number of bees, which is indicative of health.

He warns those not experienced in bee-keeping against complicated patent hives. The management of the old-fashioned box hive is simple and understood by all. There is much hope however, that the patent of Mr. Underhill, of New York, will be so perfected as to be a real improvement. Mr. Harbison's hive has many advantages, which were given in detail.

The hive should be so constructed as to be protected from the cold north and west winds. It should not be near a body of water, as bees which are heavily laden are thus often drowned. The hive need not be more than a few inches from the ground.

For hiving bees, he described an easily-constructed contrivance, which dispenses with the necessity of climbing to fearful heights where the bees may have alighted.

The methods of obtaining the honey without destroying the bees and injuring the honey, are quite simple and desirable.

ELEVENTH DAY.—Feb. 13th, 1860.

To-day Mr. Tucker gave his second lecture on English Agriculture, to an audience larger than usual.

Mr. Tucker's lecture was a continuation of his subject of yesterday, and was interesting and practical.

He remarked that, deferring until another opportunity a summary of the ground already covered, he would endeavor to describe briefly one or two extensive farms against the management of which it was less likely that a charge could be brought of any "higher farming" than was consistent with profit, or within the reach of others similarly situated. Undoubtedly there was bad farming in England, as well as in this country; there was, also, a small class of those whose operations were bolstered up on unusual capital, of whom Mr. Mechi would answer as an example, and who could not therefore be regarded exactly as fair specimens of the *practical man* in the present condition of English husbandry. He had enjoyed the opportunity, however, of visiting several who might justly rank as such, and could only regret that the necessities of the case then compelled the entire omission of much in which an interest would be felt by practical farmers in this country, and the very imperfect survey of the instances to which time allowed an allusion.

"Butley Abbey," and one or two other farms, altogether including 3,000 acres in the county of Suffolk, occupied by Mr. Thomas Crisp, together with the operations upon it, were first considered. A description was given of the sheep-walks, and the system of sheep-husbandry practised. The "four-course" system is generally adhered to, but a "stolen crop" of turnips is sometimes obtained—the seed drilled upon the wheat stubbles, and the roots fed off in the late autumn and succeeding spring, and the next crop in the course being mangolds. The quantity of mangolds grown is increasing, compared with turnips, so far as his observations extended in Great Britain.

The sheep of that part of England are prolific mothers and good milkers, and are consequently in demand. Mr. C. had a flock of about 2,000 breeding ewes, with which he puts a Leicester or Southdown "tup." The lambs it is his practice to sell, the autumn after they are one year old, or indeed any time during that season according to circumstances; and the price received for them varies with age and quality, from $7.50 all the way up to $15 per head. The lambs are dropped about March, and when they are ready to wean after harvest, are put out upon the stubbles to eat the "seeds" that were sown in the spring, and at night perhaps folded upon a turnip field as soon as the latter is ready. But Mr. C. keeps a great many sheep out a-boarding, as we might express it; that is, there are many smaller farmers, who do not have the means of keeping a large flock the year round, and who are glad to take in those of their neighbors both upon their stubbles and to eat their turnips. For the lambs thus sent out upon stubbles on other farms, about 3 cents a head per week is paid. The price paid for turnip land is in the neighborhood of 6 cents a week for each head, though it varies with the character of the crop, &c.; when it does not exceed this price, Mr. C. considers that there is room for profit to the owner of the sheep. Sometimes he has flocks at a distance of 50 miles or even more, and a great advantage of this method to the small farmer, arises from the fact, that while the few sheep he would want to keep might be all winter in eating his turnips off, if 500 or 600 come upon his fields at once, they are all cleared by Christmas and ready for plowing.

In a train on the way into Lincolnshire Mr. Tucker met a farmer of that county who had sheared, the preceding spring, 1,200 sheep, a large number for a farm of eight hundred and fifty acres. He had mentioned also the practice which some of us have advocated and others decried so strongly—that of spreading the manure upon the wheat-lands some time before plowing up the stubble of the clover crop, and permitting it to remain in exposure; a method of which he was strongly in favor, and which has been long and successfully practised by John Johnston and others in this country.

The next visit spoken of was at Aylesby, also in Lincolnshire, the residence of Mr. Torr, a noted Shorthorn breeder, and extensive farmer. He cultivates about 2,100 acres, mostly of "fen" land, although not of that lower kind requiring drainage by steam or wind power. He was an ardent believer in deep drainage, and had spent during the year before not less than $10,000 for oil-cake, guano, and artificial manures. He had 500 acres in wheat, 250 in barley, 100 in oats, 415 in mangolds and turnips, 335 in artificial, and the remainder in permanent grass. He annually shears about 2,000 sheep, and has an annual show and "letting" of breeding "tups." His average crop of wheat is nearly 40 bushels per acre (say 36 to 38), bad years with good, and he thought that the whole county would be from 30 to 32.

Some remarks followed upon the expense incurred by English farmers to remove *quack*, *couch*, or *twitch* grass, as it is variously called, and the presence of which is considered inimical to any crop. A description of the mode of plowing advocated by Mr. Melvin, an intelligent gentlemen and farmer in Mid-Lothian, then succeeded. The important points in the construction of the plow were such a medium length in the mold-board as not to break up the furrow-slice too much, as it will if it is too short, and, on the other hand, not to polish off its exposed surface too smoothly, instead of leaving it so rent and torn that the elements will act properly in the disintegration of its particles. Above all, however, a plow should turn a clean furrow, for if the earth anywhere adheres to the mold-board, the friction wastes power, the furrow is imperfectly turned, weeds are not covered in, and the old surface is not well turned under.

On the Tay, opposite the noted Carse of Gowrie, he had found a seven-year course of rotation in vogue, viz.: 1, wheat; 2, barley; 3, grass; 4, oats; 5, potatoes, or beans; 6, wheat; and lastly, turnips. The soil is so stiff that a very good drain is made by simply digging a channel of several inches' depth with a shoulder on each side of it, in the bottom of the drain, and covering it (the channel so formed) with flat stones; this being nothing else than the "shoulder drain" already described by Judge French. Grain appeared to be more generally sown broadcast than drilled in Scotland. The women were at that time at work reaping; five women with sickles, to one man binding, and the whole *gang* paid 12 shillings sterling, say $3 per acre.

One of the last visits before leaving England had been made in quite an opposite direction, namely, among the hop-gardens of Kent. As there was one lecture in the course devoted to that plant, he gave a few facts in regard to the general system of farming pursued. The farm he had seen was one of two hundred and seventy acres, and a vineyard rotation was practised. For example: 1, turnips; 2, barley, or oats; 3, wurtzel; 4, wheat; 5, red clover; 6, wheat; 7, barley, or oats; 8, beans, or peas; and 9, wheat—thus securing five white crops, three of them wheat, to four green crops. To take this rotation from the beginning, the turnip crop will have been preceded by wheat; after that was harvested, a kind of plow or cultivator, called a broadshare, was passed over the land, a flat point eighteen inches wide being carried about three inches below the surface, not turning over the ground at all, but cutting off the roots, and killing the weeds. By this operation and the subsequent harrowing, the ground is so stirred that the seeds of noxious plants, as well as those self-sown by the last crop, will vegetate. Immediately after the broadshare, the harrow is twice used to free the ground from the stubble, which is gathered in rows every fifteen or twenty rods, according to quantity, and if thought worth the labor, or in default of straw enough, this is carried to the yards, to be trodden into manure; otherwise it is burnt. A second plowing takes place, if possible, before the middle of October, say eight inches deep, burying any vegetation that has started, and throwing the soil into furrows as rough as possible, in order that the frost may act upon it; for the rougher and the larger lumps in which it lies, the better will a spontaneous disintegration be effected during winter. The next process is a plowing the last of March or the first of April, after which the land is harrowed twice, and rolled. The second spring plowing is done with the broadshare, and after another harrowing and rolling, the manure is carted out and spread, and plowed in six or seven inches deep. Then there is another harrowing and rolling, and the land lies about a fortnight, when, if the weather is dry, the broadshare may be once more employed. Swede turnips are sown about the first week in July, and white turnips about the third week—about half and half of each being grown. If mangolds was the crop, the preparation of the land for it would be similar, except that one plowing would be omitted, as the seed is sown the second week in May.

The lecture was concluded with an extended and detailed statement of the notes gathered at Burley Hall, the residence of Thomas Horsfall, Esq., whose experiments in stock-feeding and in dairy management have attracted so wide attention. A minute account was given of his fields, meadows, and pasture, of his farm buildings, his dairy room, &c. Upon not quite sixty acres of land he was keeping the following stock:

Heifers and Bullocks..........	21	Old Sheep....................	64
Milch Cows.................	20	Lambs......................	106

Likewise, 4 pigs, 2 horses and a pony.

Being a total, small cattle and large, of 218 head.

The interest of this farm is chiefly in its stock and in its grass fields. The sheep (ewes) Mr. Horsfall generally purchases in October, to the number of say fifty; paying about $11 25 apiece. Fifty-nine, a cross of the Cheviot male on Leicester ewes, procured in the autumn of 1858, had brought him the one hundred and six lambs he had to sell in 1859. These were sold before the end of July, the purchaser taking any before if he chose, at about $6 each. The ewes are fattened and sold in the fall, fetching about $12 25 each, being $1 advance on the purchase money, she having brought him during his possession of her a lamb and a fleece besides. The bullocks fattened on the farm are bought in April or May, grazed through the summer, stall-fed in the early autumn, and sold in November.

Cows are generally bought just after the second calving, though a good cow is bought at three or four years old and at any season. They are milked from three to four years; though only the longer period when their good qualities seem to warrant it. They go dry from two to three months in the year, and by skill in selection they average twenty quarts per day, when fresh. The breed preferred is a cross, half Shorthorn and half Highland, a sort plenty in that vicinity. He generally pays about $75 per head. The cows are kept in good order, Mr. Horsfall maintaining that his success depends on this, and that at the end of a cow's sixth year, when her milking qualities begin to fail, he has an animal ready, by a little "finishing," for the butcher, thus getting both the milk-man's and the stall-feeder's profit out of the same animal. But it is the management of the pasture and meadow-lands which claims our special attention. Fourteen acres of meadow can pasture twenty cows and twenty-four sheep, with a little assistance, till the middle of October. Another lot of twenty acres, every foot of which the cattle will eat, has usually supported one bullock and a sheep and a half to each acre. To these pastures the stock is not admitted until the grass is well up, this being a security against drouth. Previous to this they graze in the mowing lands, which are cut down close by them, but which produce at the end of June two tons and a half to the acre, besides a second crop, or after-math.

The best pasture is a deep alluvial loam, but the meadow, an irrigated one, is a thin soil, and a stony clay. The irrigating water is the sewerage of the village of Burley flowing into a small brook which is turned on to the meadow at the highest point, and conducted in channels to all parts of it. It runs on during the winter; is turned off in the spring to allow of grazing; turned on again to start the grass, then off to harvest it, and on again to start the second crop. The "little assistance" which the pastures have in supporting these animals, is a small quantity of cooked food, when the feed begins to fail in August, given to the cows. They are stabled at night, and receive a "foddering" of grass often cut from the pasture itself, where the droppings of the animals have caused a growth too rank to be eaten in the field. In the hot season the animals are stabled during the day, and let out to graze in the night. Of grasses, Mr. Horsfall prefers the *poas* and *festucas*—what is there called meadow grass, being the best known variety of the former genus. All his lands are drained; the lines of tile running eight yards apart and three to four feet deep; the latter depth being preferred.

A description was given of one of the stables for feeding, including the measurements made upon the spot. The roof is of slate, with a thatch underneath. The stalls are about three feet nine inches wide, and the cattle fastened by sliding rings and stanchions about a foot back from the manger. At the upper part of the stall lies a cocoanut mat, about three feet square, with straw underneath, the whole fastened securely down. Behind this mat, the only bedding the animal has is a grate, allowing the passage of the manure into a tank underneath, which tank is accessible from the outside of the building.

The manure removed from this tank is mixed with the scrapings of the road and the cleanings of the ditches, and applied to the meadows at the rate of a dozen loads to the acre, just previous to a shower. There being no straw or coarse material, it is immediately washed in. The time of manuring the meadows is as soon after mowing as the weather is suitable, and for the pastures the winter season. Liquid manure is also applied to the spots of the pasture where the grass is coarse or wiry, and also to spots comparatively bare. Three or four doses are given during the winter, but if there is an excess of liquid manure, it is poured into the stream which irrigates the meadows. The manure from an animal, if properly cared for, is estimated on this farm at $25 per year. In regard to the use of liquid manure, Mr. Horsfall disagrees with Dr. Voelcker's theory, published in the Royal Agricultural Society's Journal, that "soils containing a fair proportion of clay, especially stiff clay soils," are not benefited by its application. The experience of Mr. H. is to the contrary. Dr. V. also advocates diluting liquid manures; Mr. H. objects, and thinks the former draws his conclusions too exclusively from the Flemish farmers of Belgium.

The food for winter feeding is steamed, the rations for each cow being—rape-cake, 5 pounds; bran, 1½ pounds; malt combs, 3½ pounds; Indian meal, 1 pound; with straw, cut to ½ inch in length, 10 to 12 pounds. This mixture is dampened, care being taken in this particular, as the laxative qualities depend on the amount of moisture it contains, and then steamed one hour. The materials are changed according to the price. The weekly cost of this cooking is four cents per head,—one man, with a little help in milking, having the charge of twenty cows. The price at which the milk is sold is four cents per quart, and as the demand does not always come up to the supply, the remainder is used for butter-making.

Everything he had seen of Mr. Horsfall's practice, in fine, could not be regarded as less instructive than his essays have been, and the two consulted together, furnish facts of universal value, and hints as well capable of being turned to good account here as in England.

At 7½ in the evening, Hon. Josiah Quincy, Jr., of Boston, gave a very fine lecture upon "The Profits of Farming and the Position of the Farmer." To 20,000 lawyers, and 100,000 merchants in our country in 1840, there were 2,400,000 farmers, and the number would not fall short of 3,000,000 now. The first question always asked about farming is, Will it pay? Will the returns for all my labor be remunerative? He then proceeded to consider the gentlemen farmers who work for amusement, as not coming properly within the category. And there the contrast was strikingly drawn between the English or Continental farmer, whose rents and taxes are enormous, and who farm at the worst advantage, and the free noble American cultivator of the soil. The first error in New England is in keeping the accounts too loosely. Not one farmer in ten knows what it costs to raise a cow or a crop of corn. In England an exact account is kept with every field. Another error is the want of economy in modes of farming. Two things are required for successful farming—intelligence and capital. "Experiment," says Liebig, "is a question put to nature, and the result is her answer." Two things, labor and manure, are also necessary for a large return. It has been said that the requisites for success are three: First, manure; second, *manure;* and third, MANURE. The real profits of the farm arise from the circulating capital. An English farmer who had just leased a farm for $8,000, spent $50,000 for stock, tools, seeds, &c. A farmer can't afford to own bank stock, for he wants the money in his business. All the manure that is requisite should be the product of the farm. Dr. Dana, of Lowell, ascertained that each cow gives, when housed, seven cords of manure annually, and when mixed with two cords swamp muck or peat to one of manure, would give 21 cords of dressing equal to that of the barn-yard. It is worth from $5 to $8 per cord. The milk the same cow would give would be worth at the outside $65.76, while the manure would be worth from $105 to $168, and this is usually lost. Mr. Quincy then drew a parallel between the wealthy merchant and the successful farmer, making the average life of the latter double that of the former, and he also carrying out more fully the designs of the Creator, and finding health and happiness the truer recompense. He closed with an eloquent tribute to the worth of the American Farmer, and his value now, as in Revolutionary days, to our common republic.

TWELFTH DAY.—Feb. 14, 1860.

Judge Henry F. French, of New Hampshire, told us, on being first introduced to the Convention, that he was not an orator; but his audiences of yesterday and to-day are, if I may judge from their expressions at the close of the two discourses, convinced that he is possessed of the eloquence of facts, more useful to us than the other glittering qualification. He commenced this morning by saying a good thing boldly, viz.: that open ditches obstruct good husbandry, a fact which the opponents to covered drains would do well to remember. Open ditches occupy much land needlessly; they cause constant turning at headlands; their influence on the area of soil is not uniform, as the parts nearest them are dried while the rest is left as wet as ever; in heavy rains not only is much soil washed into them, but, along with it, manure that at labor and expense has been applied; their banks washing away, the bottoms soon get filled up, and require frequent cleaning out; and their sides and boundary strips afford a refuge to weeds, and a home to rats, mice, and other vermin. Sometimes, as "headers" to cut off the inflow of water to a field, they may be of use; and again, on very level land, a great canal-like ditch may be employed, in lieu of a natural water-course, to receive the drainage of a farm; but these are the exceptions to a general rule. The various kinds of drains were in turn described, the lecturer observing that there might be circumstances where tiles could not be had, and thence these several substitutes could be tolerated as makeshifts. In brush drains, the durability of the material depends not so much upon its keeping nature as on the physical and other character of the soil. Thus, he had known an instance of white-birch, which one would think would decay in a year, having remained in a brush drain for six years almost as fresh as when cut.

The reason for its preservation was, that it had been submerged in water continually. Into brush-drains soil very easily falls, and soon here and there the superincumbent mass caves in, sometimes to such an extent that a wagon-load of dirt is required to fill the sinks; mice and moles work into them, too, and at best they are poor concerns. The mole-plowing now practised on Western prairies is, for a new country where land is so cheap, and where a sticky clay sub-soil underlies whole districts, a tolerably good plan. It has been known and practised in England since almost the time of Methuseleh. Major Dickinson of Steuben county, New York, has gotten up one of these ancient mole-plows, and dubbed it "the Shanghae." Drains are made in some "wooden countries," by laying two stout poles at bottom and one on them. In Scotland they have in some benighted sections a "shoulder" drain, which consists in digging down, say 18 inches wide, to a certain depth, and then cutting the rest of the way down only one-third as wide; thus making a narrow box drain in the ground on the shoulders of which inverted stiff sods are laid as a covering, and the soil filled up to the surface upon them. Stone drains he esteems next in utility to tiles, but there is great choice in their construction. The best way of all is to set up one course of slab stones perpendicularly against the right bank, and then leaning other stones against them, making a drain shaped like a single-pitch shed-roof. If the stones are delivered to a farmer at the edge of his ditches, they are still dearer for his use than tile drains, even when he has to pay $10 or $12 per 1,000 for tile. The mere cost of excavating and hauling bowlders for drains is very large, and after all, their function is unsatisfactory. The reason why all these kinds of drains have been stoutly upheld by their users is, that any drain, however poor, is far better than none; crops are increased, tillage facilitated, and the pleased experimenter, perhaps not willing to look for a better method than the one he has employed, thinks there is nothing in the world so good. Tile drains, then, we are told, are the best. Of the several kinds of tile, the pipe kind is to be preferred. No tiles are burned, without warping and shrinking; now the ends should be well fitted together, and no kind but pipe-tile can be turned over to make good fits, one with another, and still be right side up. This is the objection to the sole-tile, made at Albany and elsewhere, and largely employed. They must be set sole down, and if the lot purchased be much warped, a straight water-course cannot be insured, and the drain is correspondingly unreliable. The objection to "horseshoe" tile is, that in a soft bottom its narrow sides sink so as to render the drain sometimes useless; besides which, they, having a heavy weight to bear upon an unarched bottom, are liable to split lengthwise through the back; and, further, the stream of water spread over a flat surface cannot run so rapidly, and is less able to sweep away obstructions, as when the same volume is condensed into tubular form, narrowed at the bottom. Thinking that water could not get into the close-fitted and close-textured tiles, many in Scotland, in former times, put a foot or so of small stones over their tile, and soil upon that—a foolish and expensive process this, for there is no trouble to get water into your insignificant-looking drains—it takes care of that itself; the trouble has been to account for its wonderful inpouring through such small orifices. Parkes, the great English drainer, states, after experiments, that only $\frac{1}{500}$ of the water gets through the pores of the tile; the balance is admitted through the joints. English farmers make their ditches a foot wide at top, four inches at the bottom, and with an appropriate tool, scoop out a little round trough in which to lay their pipes. The soil is then packed upon them, without further trouble or anxiety as to the result. Drains well laid last more than fifty years. A half century is the time counted upon by the English land drainage companies, at the end of which the whole amount of their loans to the farmer is to be paid in. Water enters tile-drains at bottom, not at top; for the same reason that if you pour water into a cask of sand, with holes made in the sides at several heights, the lowest hole will discharge first, and the top one last. The capacity of pipe-tile is in proportion to the squares of their diameter: Thus, if an inch tile will carry one inch of water, a two-inch will carry four inches, a three-inch nine, and so on. Inch tiles, therefore, although perhaps large enough to hold all the water that we would discharge from our fields, are practically not large enough, for they become filled at say half way down the slope, and of course all the ground they pass through after that might as well have no tiles beneath it. A two-inch bore is the smallest Judge French would recommend for general use, and although previously a friend to smaller sizes, I feel convinced of the justness of his arguments, and shall hereafter recommend and use accordingly. Laterals should be jointed into the mains, *pointing down stream*, and enter the mains near the top; by this plan a good fall and unimpeded discharge are insured. In respect to the minimum of fall consistent with good function of tile drains, the lecturer stated that one inch fall in each rod of length was ample; three inches to the 100 feet was a fair proportion, but then the tiles should be larger; and so on to the end of the calculation.

Before the morning lecture, a discussion was held at the Temple, as usual, in which any person present was at liberty to participate.

Mr. Quincy alluded to the advantages of the soiling system —his pet subject—in doing away with interior fences on a farm. These, said he, are a great nuisance, besides taking up valuable space; they hinder plowing, raking, tedding, and other operations of farming by horse-power. Tedding by horse power is something new in this country, though practised in England extensively. The tedder is a cylinder, revolving on an axle supported by two wheels, like a Delano horse-rake. This cylinder revolves with rapidity, and is furnished with teeth, which pick up the grass and flirt it off in a shower behind the machine. It will do the work of ten or twelve of those Irish gentlemen who pick up and turn over every lick of hay as though they were fearful of breaking it. The horse-fork is also a labor-saving instrument; it also avoids the very disagreeable work of unloading hay in a hot day and in a close barn.

But the great advantage of the soiling system is, it saves manure. It economizes food, it is true, and keeps cattle in better condition, but its chief excellence consists in the amount of manure it will make. The solid manure from each animal, kept up the year round, will average three and a half cords a year; this, with the liquid manure composted, as it ought to be, with muck, will make twenty cords, of a value equal to that usually carted out from a farmer's barn-yard. Four or five hundred cords of muck are annually dug out on his farm, and left exposed to the weather in winter. This is used, when dry, to put behind the cattle in a trench made for the purpose. After it is saturated, it is removed to a cellar below, where it would be worked over by the pigs were it not too miry for them to work in. This makes, in the course of a year, a vast pile of manure; so much, indeed, as to remind one of the Augean stables of antiquity, and to seem to require the services of a second Hercules for its removal. The soiling system is almost universally adopted in Europe; it may not be practicable here, except on a large scale, though almost every farmer can use it to help him through the drouths of our summers. In case the drouth does not come, his crops, which he has planted for soiling, can be cut and made into fodder for winter use. The supply of milk, under the soiling system, is much more regular, because the cows are regularly fed, regularly attended, and fed always with the same kind of food. For soiling, sow winter rye, to be cut early in the spring, and in the spring sow oats or barley every ten days, so as to have a regular supply in just the right season,—that is, when the plant is in its milk. Indian corn is also a good crop for later use.

Mr. Quincy here spoke of seeding down land to grass. He

had found it a good plan to break up a meadow after haying: manure well on the turned soil, and sow grass seed only. The next season he had cut from two to two and a half tons to the acre, where the previous season he had cut almost nothing.

Question—Do you buy any manure?—No; but I buy cotton seed cake to feed my cows. This is, at present prices, the most valuable feed to be had,—a ton of it being worth, at the chemist's estimate, three tons of hay. It is now worth $27 per ton in Boston. Linseed cake is also valuable, and English farmers wonder how American farmers will let it be exported in such vast quantities as it is.

Judge French asked Mr. Quincy if he fed roots.—No; Linseed cake and hay is the sole food—three pounds of the former per day, with cut hay.

Mr. Bartlett asked what Mr. Quincy's advice would be to young farmers here, in regard to going west—alluding to Mr. Q.'s travels there.

Answer—If a young man will be content with the same living here that he will be obliged to put up with there, he can make money here as well as there. They have no idea of what decent living is there. Then, too, there is no society at the west—no schools, fit to be called such—no aristocracy. There is a perfect equality there; your Irish gentleman who curries your horse feels himself to be your equal, and not unfrequently your superior. Civilization is in an embryo state, society not yet having advanced to that perfection which we see at the east.

Question—How does soiling affect breeding?

Answer—I do not think it prejudical. I am not a stock raiser myself, but farm merely for the profit. I buy my cows in Vermont and New Hampshire, though sometimes I raise a likely heifer calf. There are cows in my stable whose maternal ancestors have been there for eight or ten generations past.

Mr. Tucker asked if ventilation was attended to.—Yes, and with great care.

Question—Is lucerne grown on your farm ?—It is difficult to make it yield a good crop, and I don't consider it profitable.

Mr. Quincy was here obliged to leave the Convention, and the subject of root crops was introduced by Judge FRENCH, and an animated debate held on this topic until the lecture hour arrived.

THIRTEENTH DAY.—FEB. 15, 1860.

Prof. BREWER opened his Tobacco lecture yesterday with a rapid sketch of the history of the imperial weed, and referred to the pains and penalties which attended its use under successive sovereigns. The chemical composition of the plant is very remarkable, and worthy of serious study by present and prospective growers. Nicotine, the deadly principle to which all the ill effects of tobacco are due, is, as every one knows, a deadly poison. Besides this, the plant contains a number of acids, resins, and volatile oils. The strength of tobacco is determined by the quantity of nicotine; the flavor by the oils and resins. The ash is of all the most important to the farmer, for this is made up from his available plant food—in other words, from his farm capital. The oils, resins, and acids come from the air, and hence cost us nothing. Take a given quantity of tobacco and burn it to ashes, and we find that the proportion is enormous. The roots give two to fourteen per cent. of ash, the stems dried sixteen, and the leaves seventeen to twenty-four per cent. As the leaves are the great bulk of the crop, the robbery of the soil is correspondingly great. One thousand pounds of tobacco takes an average of two hundred pounds of ash; and two thousand pounds, which may be regarded as a large crop, four hundred pounds of ash. Now, a crop of wheat of thirty bushels to the acre takes but thirty-six pounds of ash from our farm. In other words, it would require *eleven crops* of wheat to do as much injury as a single crop of tobacco. The composition of the ash is variable, in some districts one of the leading ingredients being replaced by some other. In an average of samples tested by Prof. Brewer, potash salts formed a third part of their weight, and seventy-five to eighty per cent. of the soluble portion. Soda exists in but a small quantity. Sometimes the potash is replaced by lime. Thus in France, along the river Garonne, the tobacco has this peculiarity, and is noted for its mildness. In American tobacco, the potash salts predominate, and most so in the stronger kinds, which grow on new soil. A study of the census will show us, that in any tobacco district, the production starting at nothing, mounts rapidly to a maximum, turns the corner, and never regains its higher figures. The reason is, that land can only bear maximum crops of tobacco for a short time, and once the decline comes on, no power on earth can restore its fruitfulness. By high manuring, we can, with other crops, actually improve the fertility of our farms, or at any rate, guard against impoverishment. With tobacco, if we manure highly, we may for a time avert the *dies iræ*, so far as bulk of crop is concerned, but only at a sacrifice of quality so great as to destroy our profits. New crops have coarse quality of structure, and rankness of flavor; while, *per contra,* the tobacco of finer brands is gotten from lands long cultivated. A thin leaf, with small pliant veins, is most esteemed, and of this character is the tobacco of Holland and Connecticut. The season of growth is ordinarily crowded into forty days, and the larger portion of the soluble salts must be at this headlong speed, supplied to the spongioles. The crop is so tender, that of all those we cultivate, it is the most subject to destruction by hail. In Germany there are "Hail Insurance" companies on the mutual plan. It is a notorious fact that hail-storms extend over very limited areas at a time, and hence the farmers of a whole country uniting in small annual payments toward a mutual fund, it will be seen that even the most disastrous hail-ravages could easily be recompensed, without fear of extinguishing the grand capital. In considering the advantages and disadvantages of tobacco-culture, Prof. Brewer thus stated the case. The sole advantage is, that an individual may grow rich from raising it. On the other hand, a nation never will; for the one man's gain is obtained at the cost of his son and son's son; in getting his fortune he has taken from his children the means of future gain, like the owner of the goose that laid the golden eggs. The crop terribly exhausts the soil; it is very precarious because of weather and insect enemies; the laborers who cultivate it suffer in health; and the land, which must always be of the best quality, could be employed in raising breadstuffs to more general profit.

Mr. TUCKER's third discourse touched more generally upon the lessons which Americans may learn from the well-informed farmers of Great Britain.

Although the lectures of the succeeding week were to be devoted particularly to the subject of domestic animals, one could not pretend to speak of "English agriculture" and omit all notice of the improvements effected in English breeding, without placing himself in the position of the theatrical company which proposed to "play Hamlet," with the part of that distinguished character himself left out. The subject might be viewed in two different ways—with the eye of the farmer, or with that of the breeder—a distinction of more importance than might be at first supposed.

After a review of the breeds of cattle of Great Britain, it was remarked that in speaking of the most meat, in the best shape, in the least time, as constituting the highest type of excellence for the butcher, it should not be forgotten that no one breed could be fixed upon as universally superior to all others—even though there might be a "best breed," and undoubtedly there is, where every condition is of the most favorable kind for comparative development. Such conditions, however, are not within either the reach, or the inclination of all, and that may, therefore, be safely defined as the best breed, either of cattle or of any other race of animals, whose services or flesh are useful to us—which attains the greatest excellence compatible with the position it is to occupy and the treatment it is to receive. Thus, the requirements of East and West, North and South may vary widely as to details, while all might precisely coincide in the general desire to produce the heaviest flesh upon each carcase most compactly and quickly.

The importance of this point becomes apparent when we see a farmer induced to try some improved breed, and meeting with the failure due to his ill-treatment or simple neglect—a failure which he is sure to charge upon the "humbug book-farming notions" of the day. There need be no hesitation in saying that the most highly improved of foreign breeds are not adapted for the use of the majority of our farmers, and that we shall naturalize among ourselves breeds that may justly be regarded as "the best," only as we learn to appreciate and treat them better.

The question then arises, What is the true course for our farmers to take? a question which was answered by references to the observations made by the speaker abroad, and by a quotation from "Morton's Cyclopedia"—the advice derived from both being to the end that every farmer should carefully select the females from which he is to breed, no matter what their mixture of native or foreign blood, and that he should never employ a parent of the other sex which did not possess well concentrated merits that would be quite certainly imparted to his progeny. "It is here that pedigree becomes of actual money's worth to the farmer." *Concentrated* qualities in the bull are those—whatever the degree in which the particular individual possesses them—that are *hereditary* in the stock from which he springs. In selecting a bull by the eye alone, personal merits may be chosen, but the character of the progeny will very likely revert to the inferiority of its remoter parentage. It need only be suggested, whether the improved bull obtained be Devon or Shorthorn, or Hereford or Alderney—that his descent be unquestionably pure, and that a line of action once marked out be perseveringly followed—a course that could not but effect far greater results in a period comparatively short, than those who have not made the experiment will perhaps at first be ready to admit.

"Division of labor" has been strenuously insisted upon by the best English stock authorities in the business of raising breeding stock. Those who have the wealth, the leisure, and the taste necessary for this pursuit should be allowed to carry it on, while the farmer will find it his best policy to pay a fair price for a good article, rather than to run the risks of endeavoring to maintain for himself a herd of some pure and distinct breed. In the hope of obtaining some trait of superiority he does not already possess, the breeder may well pay such prices for an animal likely to beget it in his offspring, as it would be the merest folly for any farmer to expend for the worst beef that was ever contained in one skin.

The English custom of letting the services of bulls as well as rams was then described, and a brief account given of the ram-letting last summer of Jonas Webb and Mr. Sanday.

The estimates of live stock for Great Britain now show that she supports for *her whole area* the enormous number of one sheep to every two acres and a half, and one head of cattle to every nine acres and a quarter. According to the N. Y. State census of 1855, there was then one sheep to every eight eight-tenths acres (nearly), and one head of cattle to 13 and a quarter acres (not quite); but the greater weight of the English cattle and sheep over ours is probably enough considerably to increase the disproportion. It was remarked by Lavergne, and cannot fail to have been observed in the examples given of English husbandry, that it "is the English farmer's first object to keep as many sheep as possible."

Mr. Tucker conceived that the first and most prominent lesson we could learn from the farming of Great Britain was this, that by the *increased growth of meat* our first step must be taken toward an increased production of grain; or, to quote the proverbial English form in which this lesson is compressed into four words—"No cattle, no dung; no dung, no corn." In fact, whether money is apparently made or lost by feeding in England the farmers there appeared to coincide in the opinion that without it no money could be made out of anything else. A second most important lesson is, the proper and complete drainage of the soil, with reference to which an account was given of the draining and irrigating operations at Teddesley in Staffordshire, the seat of Lord Hatherton. A third lesson for us to learn consists in paying more attention to thorough tillage, including the complete clearness of the soil from weeds; and a fourth, the judicious employment under certain circumstances of artificial fertilizers and purchased food—including under these two heads those crops grown expressly for their improving effect upon the land, or for use in feeding animals, and thus indirectly in promoting the fertility of the soil. Under the head of thorough tillage, the implements of Great Britain demand our particular notice. Descriptions were given of Fowler's and of Smith's systems of steam cultivation. Mr. Bright, Lord Hatherton's very intelligent and successful manager, was employing the latter, and had said to the speaker, that he would not be without it if he were only a tenant farmer with 300 acres to cultivate. The prices of these and other implements were given, and drills, rollers, and portable engines were particularly referred to. That island, including England and Scotland, had just been compared by Mr. Morton to one immense farm, the culture of which was originally entirely done by hand; tillage of the ground, carriage of manures, sowing the seed, and three-fourths the hoeing of the crops were now done by horse-power, threshing of grain and cutting of straw by steam, while reaping, also, is now rapidly coming under the domain of the horse and plowing under that of steam.

Upon the subject of manures, Dr. Voelcker was quoted as supporting by science the lesson of practice, that "farm-yard manure is a perfect and universal manure," and that no one can base a system of improved cultivation solely upon the purchase of artificials. The fifth and last lesson of English agriculture at present noticed was the importance of more earnest and better organized effort in obtaining well-conducted experiments in carrying on scientific investigations, and in deciding that most difficult of questions, how and in what the education of farmer's sons is to be advantageously modified and advanced. Prominent among the agents of progress in English agriculture had been the Agricultural Societies; and in referring to the show last summer of the Royal Society of England, three points were alluded to as particularly striking: 1st, the extraordinary turn-out of implements, comprising 4,700 entries for some 235 exhibitors; 2d, the uniformity of excellence among the animals, as more remarkable than the number that were exhibited on the one hand, or any especial instances of wonderful merit on the other; and 3d, the character of the attendance, the amount paid for admission, and the fact that so large numbers were ready to pay it. The exertions put forth by the distinct societies were also noticed, and details given of the different exhibitions held by that of East Lothian in the course of the year, including the prizes respectively offered according to the season.

In conclusion, he could only be sensible how very small the beginning was that had been made—however long his notes might have appeared to his audience—upon the grand stores of agricultural information looked up in the practice of English farmers. He was inclined to consider it well worth some self-denial to the young American farmer to visit Great Britain before "settling down" for life—if his visit could be made in the right spirit, and judiciously arranged. In returning, he thought that the observant traveller could but bring back a better appreciation of the advantages possessed in his own land, and, however far behind our English brethren we must now be compelled in due candor to rank ourselves, if we were only certain that we were in the right path, perhaps we might still hope to overtake and outstrip them. He was inclined to believe—although there might be no statistics in support of such a statement—that a thorough English farmer, knowing our climate, and understanding all the circumstances of farming here as well as he does at home, could make agriculture here a still more profitable pursuit than he made it there,—of course supposing that he employed the same capital, and used equal, but no greater personal exertions.

To-day Prof. Brewer lectured on Hops, which he said was a crop of growing importance. In 1840 we raised 1,238,000 pounds; in 1850, 4,497,000. He traced the history of the plant, and showed that its general use can be dated only three hundred years back. England uses forty million pounds, paying to the government a duty of over a million dollars. If only the hop flowers are taken from the farm, the crop is not of so exhaustive a nature as tobacco; but still it is very much so, after all. From a ton of hops we may get 170 pounds of ash, of which potash, lime, and ammonia form principal ingredients. Liberal applications of manure are needed, and they do not affect the quality of the product, as is the case with tobacco. Beside farm-yard dung, wool, hair, bones, plaster, lime, and ashes, are all useful fertilizers. In England, the Kent and Sussex hop-growers calculate upon spending about fifty dollars per acre for special manures, in addition to what of the ordinary kind they make on the farm. With such care, they have hop plantations three hundred years old. The ground must be trenched and worked deeply. About 1,200 hills is the proper number per acre, and for each two hundred hills there should be one hill of male plants. It is better to plant in triangular form rather than square. That is to say, the hills of adjoining rows should alternate, and not be set opposite each other. When picked, the hops should be at once dried, and this is better done by passing a current of hot air over them than in placing them in a room where they get only the radiated heat from a stove. Liebig recommends exposing hops to the fumes of sulphur, as thus the *lupuline*, or active principle, may be preserved from one season to another. The practice is opposed by some, but adopted by many of the best Munich brewers. The hop crop varies from year to year to such an extent that the price is very fluctuating, and even in a single season a month may make a difference of one hundred per cent. In conclusion, the lecturer detailed the casualties to which the hop is subject, such as insects, weather, &c., and gave practical directions for its cultivation.

Judge French gave his third lecture on Draining, taking up this time the subjects of the Arrangement and the Cost of Drains. He spoke of the necessity of system, and of accurate plans. He described and illustrated on the black-board the methods of laying out drains with reference to the shape of the field, preferring a direction up and down to a direction across, or diagonally. He spoke also of the importance of securing outlets against frogs and moles by means of gratings, and of making the outlets few and permanent. Backwater usually does no harm in drains, because it occurs only when the earth, as well as the streams, are full, and so there is a strong current through the pipes which will prevent any obstruction, as water cannot back up into pipes already full. The cost of this in this country is twice as great as it should be; two-inch tiles are sold at ten dollars or more a thousand, which is twice the cost of bricks. In England tiles cost and are sold at less than the price of bricks, and will be sold at five dollars per thousand here as soon as tile-making is understood, and there is a fair competition.

The items of the cost of drainage are, 1st. Engineering. Employ a competent engineer to get the levels, and locate the drains and make a plan, so that the drains may be readily found. 2d. Excavation, which is less for this than any other drains. 3d. The cost of tiles and freight. At thirty-three feet apart, 1,320 pipe will lay an acre, reckoning a foot to each pipe. 4th. Collars, if used. 5th. Outlets, a small but necessary item. 6th. Laying the pipes, a small cost, as a man can easily lay one hundred and sixty rods in ten hours.

The total cost of draining four feet deep, with tiles at ten dollars per thousand, was estimated at fifty cents a rod. If the excavation is but three feet deep, it will reduce the cost to thirty-three and a third cents, as it costs twice as much to excavate a ditch four feet as three feet.

The comparative cost of stone and tile drains was given; the cost of tile drains as above that of stone drains at more than twice as much, the excavation being twenty-one inches wide, and two loads of stones at twenty-five cents each, making the cost of these two items at one dollar a rod. Then add twenty-five cents per rod for laying the stones, and we have one dollar twenty-five cents per rod for stone drains, against fifty cents for tiles. Judge F. concluded with an exhortation to farmers to drain with stones if tiles cannot be procured; but not to be satisfied with their operations until they have tried tiles at four feet depth.

FOURTEENTH DAY.—Feb. 16, 1860.

Two new lecturers were introduced to us to-day, viz., Mr. John Stanton Gould, of Hudson, N. Y., and Mr. Joseph Harris, of the *Genessee Farmer.* Mr. Gould's name was made familiar to the farming public at the time when he was chairman of the famous national reaper trial of the United States Agricultural Society. His lecture was not only replete with interesting facts and practical suggestions, but adorned with those graces of scholarship he knows so well how to employ.

After an allusion to the æsthetic character of the grasses, their economical relations were adverted to. Providence has attested their importance by the provision it has made for their diffusion and preservation. While other plants, such as the fig, orange, and grape, can only be successfully cultivated within narrow belts of latitude, the grasses extend over the whole globe. Very curious and various provisions are made for the diffusion of the seeds; many of them are furnished with creeping roots. They are not, like other plants, injured by the laceration of their herbage. One-sixth of all the plants on the globe belong to this family—230 genera, including 3,000 species, are already known, and new species are constantly presenting themselves. Six-tenths of the cultivated area of New York is devoted to the growth of grass, and the annual value of the crop is $60,000,000. In the six New England States its annual value is $6,000,000. In the United States, $300,000,000. If we succeed in making two blades of grass grow where but one grew before, we increase our annual income $300,000,000.

It was argued that we might easily double our production of grass, if we would set vigorously at work to accomplish it. The average production of New York is 96 tons of hay to the 100 acres; but the average production of King's county is 160 tons to the 100 acres. This result is wholly due to the skill of the farmers, as its natural soil is far below the average of the State in richness. If the same skill were exerted in other counties, the same result will follow. Another cause of the diminution of grass is the prevalence of weeds; at present nearly one-third of the plants in our meadows are weeds.

Much ignorance exists among farmers; very few know the names of the grasses growing on their farms, nor can they distinguish one from another. They know little or nothing of the comparative nutritive values of the different species, nor of the soils best adapted to them; nor of the special purposes to which they are applicable. It was alleged that chemistry can never, by itself, furnish a safe and reliable guide to the nutritive values of the grasses, because there were frequent obstacles to the assimilation by the animal of the nourishment contained in the grasses; thus, *Phragmites communis* (common reed grass) had a coating of silica so thick that it would cut the stomachs of animals; other species had sharp spines, which deterred animals from eating it; others combined unwholesome elements in their nutriment;—hence, whatever nourishment might be contained in these was quite useless to the farmer.

Much of observation and experiment is necessary before we pretend to understand the grasses. The making of artificial meadows is an art yet in its infancy. We never hear of them in England prior to A. D. 1681, nor in this country until about A. D. 1720. The attention of observers and experimentalists should be directed to the following points:

I. The special use of each of the 3,000 species of grass.

II. The absolute and comparative values of each species should be ascertained by chemical analysis and practical tests.

III. The adaptation of each species to different soils, climate, and circumstances.

IV. The period of its growth when it contains the greatest amount of those properties on which its value chiefly depends.

V. The kind of culture and the manures best adapted to stimulate its growth and to increase its valuable properties.

VI. The time of flowering of each species, and the time when it ripens its seed.

VII. The species of insects which prey upon it, and the best modes of preventing their ravages.

VIII. The best and most economical means of curing and preserving each species of grass.

To enable farmers to make these observations, they were advised to study botany;—and the remainder of the lecture was occupied in describing the parts of the grass which are mainly resorted to in order to establish the distinctions of species. Some of these descriptions are peculiarly valuable, because not given in any work on botany which I can now recall. The leaves consist of the following parts:—(*a*) The *Sheath*, which represents the petiole or leaf-stalk of other plants; (*b*) the *Ligule*, or tongue; (*c*) the *Lamina*, blade or flat part of the leaf, that which in popular language is called the leaf. (*a*) The sheath is the foot-stalk of the leaf. The whole length of it, which is variable, is folded around the stalk (culm), from which it can be loosened by unwinding, without fracture,—a circumstance which serves to distinguish the grasses from the sedges. (*b*) The ligule, or tongue. At the point where the sheath ends and the blade begins, occurs a thin and usually white semi-transparent membrane, termed the *ligule*. As the botanical works barely describe this, and still perplex us with constant allusions to this and other parts of which we have about as little knowledge as of the Choctaw alphabet, it is well to remark that this *ligule* is said to be *entire* when it has no segments; *bifid* when it is divided at the apex into two parts; *lacerated* when it appears as if torn on the margin; *ciliated* when the margin is set with short, projecting hairs; *truncated* when the upper part terminates in a transverse line; *acute* when it has a short, sharp point; and *accuminated* when it has a long, projecting point. It has great value in enabling us frequently to distinguish between two grasses otherwise very similar in appearance, but of widely different nutritive value. Speaking of the area under the grasses in European countries, Mr. Gould made a forcible illustration of his subject by comparing the aggregate products in forage and cereal crops in France and England. France has fifty-three per cent. of her cultivated area under cereals, while England has but twenty-five per cent. But, on the other hand, England produces five and one-ninth bushels of grain for every individual of her population annually, while France produces only five and a half bushels. Thus, with less than half of the proportionate area under cultivation, England produces within seven-eighteenths of a bushel per head of what France does. This she accomplishes solely by means of the manure furnished by her grass lands. Every acre of English grain-land receives the manure from three acres of grass-land, while in France the manure for each acre of grass-land is spread over two and a half acres of grain-land! This tells the whole story; shall we profit by the lesson?

Judge French, of New Hampshire, gave his last lecture on drainage this afternoon, much to the regret of the audience, if I may judge by the triple rounds of applause by which he was honored on taking his leave of us with a kindly expression of good-will. He commenced by reading an extract from a letter of Governor Hammond, of South Carolina, to Levi Bartlett, recently received. The testimony of the distinguished Senator is so directly in support of thorough drainage that I must give it to you. He says:—

"Of my agricultural affairs, I can only say a few words. The last years have been, in my immediate neighborhood, average crop years, the last more than average. Yet with me, owing to my absence, as far as my corn was concerned they were not near as productive as 1857. My corn is mainly grown on the 1,500 acres of inland swamp I have reclaimed, which averaged me, in 1857, about fifty bushels per acre, in 1858, about thirty bushels, and in 1859 about twenty. This looked like exhaustion; but I know it was not so. I was satisfied, from former experience, that in my absence the ditches had not been thoroughly cleared and kept clean. Before I left home, in December, I had the matter fully tested, and found that my six-feet ditches were three to four feet deep, and all others in proportion. Such was the carelessness and malfeasance of those I left in charge. I inaugurated new officers, and if next year is as favorable as the last, will expect to average seventy bushels per acre on these lands."

This very 1,500-acre corn-field I went through in 1857, and can fully corroborate what the Governor says about his large yield, and the depth of his drains. In fact, his great outside drains looked more like canals than anything else, and were, at the time of my visit, abundantly filled with water. Two acres, if I recollect aright, of this corn-field measured ninety-eight bushels each, and the plantation crop amounted, in the aggregate, to about 56,000 bushels. This was raised on a swamp, just like many thousand other acres in South Carolina, but rendered thus fertile by open ditching. Governor Hammond's experience goes to corroborate what yesterday Judge French said against open ditches. In one season only, because of neglect to clean them out, the ditches filled up, so that on the 1,500 acres the crop was shortened 30,000 bushels, and in one year more a further loss of 15,000 bushels was experienced. Let things go on at this ratio, and in 1863 Mr. Hammond might as well save his seed, for he would get no crop at all.

Judge French adverted to the fact that plant roots cannot descend into soil filled with stagnant water, for it has the same deleterious effect upon them as does holy water upon a certain unmentionable gentleman of a sable hue. All plants need loosely-packed soil, and some of them a great depth of it. The downward travel of roots he proved by the observations of Mechi, Cobbett, Downing, and others. Jethro Tull's ancient doctrine, that by extreme comminution of the soil we will furnish abundant food to plants without needing to care much for manures, although obsolete for many years, is of late coming into vogue again; and we certainly cannot work up our heavy soils as we should, unless we draw off at the bottom the excess water, which renders them sticky and tenacious. Evaporating it at the top will certainly not avail, for from a wet soil the more we have evaporated, the colder we get it, and hence the less fertile; for plants like warmth and plenty of air, as well as moisture. The several advantages which follow thorough drainage were severally adduced, and very clearly and agreeably explained by the Judge, who has a pleasant conversational way with him that interests one vastly. In England it has been found that draining makes twenty-five per cent. difference in the amount of work which animals can perform on a farm in a given time. That is to say, three horses will do as much plowing on a drained farm, as can four on one undrained, for their strength is correspondingly less taxed.

The lecture by Mr. Joseph Harris, of the *Genessee Farmer*, was not only replete with practical hints for the cultivation of the cereals, but contained, also, a full exposition of the chemical laws to which the farmer must pay attention if he would secure maximum crops. The original newspaper report of the lecture was necessarily very meagre, and I substitute, in its place, some extracts taken from the MS. itself, which has been kindly placed at my disposal for this purpose by Professor Porter.

The great aim of the wheat-grower in nearly all sections is to get wheat early. In western New York, if we could get wheat into bloom ten days earlier, we could escape that terrible insect-pest, the midge. It is this insect, and not, as has been often stated, the exhaustion of the soil of phosphates, that has caused the deterioration of our wheat product. The injury from rust, or mildew, another great drawback to profitable wheat culture, would also be greatly mitigated by earlier maturity. Now, there is no one thing that will do so much to accomplish this as underdraining. Stagnant water is not only injurious to the growth of wheat, but it renders the soil cold and retards the ripening of the grain. It has been found, by actual experiment, that a soil which needs draining is from 10° to 15° colder than the same soil after it has been underdrained. In our late, cold springs this would be an immense advantage. Having the soil underdrained, the next thing is to prepare and enrich it for the crop.

The introduction of turnip culture and drill husbandry into England banished summer-fallows from all but the heaviest clay soils. There was good reason for this: The turnips required and received extra cultivation. As soon as the wheat crop is harvested, the land is scarified and plowed in the autumn, and two or three times in the spring, and rolled, and harrowed, and scarified till it is as free from weeds and as mellow as an ash-heap; then the turnips are sown in drills from 2 feet to 2½ feet apart. The plants are singled out by hand-hoes in the rows, from 12 to 15 inches apart, and the horse-hoe is kept constantly going between the rows, and the hand-hoe whenever necessary. In this way the land is as effectually cleared and mellowed as if it had been summer-fallowed. Hence turnips have been appropriately termed a "fallow crop." But we have as yet no such fallow crop in America. I am aware that Indian corn is sometimes called a "fallow crop," because, like turnips, it admits the use of the horse-hoe; but it is not, strictly speaking, a fallow or renovating crop, because it impoverishes the soil of the same plant food as the wheat crop requires. So much has been said in England against summer-fallows, and these opinions have been reiterated so often by the agricultural press of this country, for the last 30 years, that there is a very general opinion that summer-fallows is unnecessary. This impression, while it may have done some good, has also done considerable harm. Farmers have neglected their summer fallows. In Western New York it has not been uncommon for some years to prepare land for wheat by simply turning under a crop of clover when in bloom, say in June, and then keeping the surface of the land clean by the use of the cultivator and harrow till the seed is sown, without any more plowing in the fall. On light soils this *may* be a good practice, but on heavy soils I think a real old-fashioned summer-fallow would be better; though I have seen excellent crops produced on heavy land by plowing in a crop of clover—the clover, besides enriching the soil, serving also to render it light. Still, I do not like the practice of plowing in clover for wheat. I believe in many cases a good summer-fallow would be much better.

Passing food through the body of an animal does not increase its ultimate fertilizing power; it adds nothing to it, but the droppings of animals are a more appropriate food for plants —at least for wheat—than the food which the animals consumed. It is contrary to the economy of nature to use plants which are capable of sustaining animal life for the purpose merely of furnishing food for other plants. For this reason, while I would earnestly recommend the extensive cultivation of clover on all wheat soils—while I would say to every farmer, "Raise your own clover seed, and sow it with an unsparing hand"—while I believe there is no crop which furnishes so much ammonia at so cheap a rate—no crop so well adapted to our climate and circumstances—no crop which has done and is now doing so much to increase the fertility of our farms, still I think it is contrary to sound theory and good practice to plow under such a large amount of matter capable of sustaining animal life, for the simple purpose of furnishing food for the following wheat crop. Fertilizing matter furnished by decayed clover is not as appropriate food for wheat as the droppings of animals living on clover. It contains too much carbonaceous matter—the very matter which animals need to keep up the heat of their bodies, and to form fat; and which, when the clover is fed to animals, is burnt out while the nitrogen remains in the form of ammonia—or in compounds which readily decompose and form ammonia. This ammonia is what we most need. It not only increases the crop, but *up to a certain* point, accelerates early maturity. (If we get too much ammonia and a moist, cloudy summer, it has an opposite effect—but there is not much danger of our getting too much ammonia.) On the other hand, the carbonaceous matter, forming four-fifths of the clover, is of little fertilizing value, and, certainly, on the majority of soils, is not needed by the wheat crop, while it has a tendency to produce too much straw, and to retard the ripening processes.

These remarks will apply, also, in some degree, to poor, strawy, leached, weathered manure. *There is not enough ammonia in a ton of such stuff as many farmers call manure to make hartshorn enough for a lady's smelling-bottle!!!* Instead of plowing in so much clover for wheat, then, let us convert it into wool and mutton, and if we can give our sheep peas, or beans, or oilcake in addition, it will tell wonderfully on the manure, and on the crops to which it is applied.

In preparing heavy land for wheat, it is still necessary, in many cases, to resort to summer-fallows. On the light soils we might take a crop of beans, planted in rows and thoroughly horse-hoed, and sow wheat afterwards. On heavier soils I have seen an excellent crop of wheat follow a crop of peas, which had been sown instead of fallowing. The great drawback to the peas is, that they are affected by the bug. But if fed out early to hogs, the bugs do not injure them materially, while they are very fattening and make rich manure. You can commence feeding them to hogs on the land, while the peas are still green. In England wheat is generally sown on a one or two-year old clover sod, the land being plowed immediately before sowing. As a general rule, this practice does not succeed here, because, for one reason, we sow a month earlier than they do in England, and a clover field plowed here the last of August is generally so dry that the seed wheat does not germinate evenly; and it is found, too, that the wheat is overrun with weeds and grass the next season. I think, however, if our land were cleared the way it should be before it is seeded to clover, and eaten down by sheep during the summer, wheat might be raised here with one plowing, as in England, especially if we used a little Peruvian guano at the time of sowing. In western New York manure is seldom applied directly to wheat; some say it is injurious. But I apprehend that, on most farms, the wheat would be very grateful for a little good, well-rotted manure, either plowed in or spread on the surface just before sowing. Wheat needs something to give it a good start in the fall, and a little well-rotted manure, not plowed in deep, would be very acceptable. A dressing of Peruvian guano, say 150 lbs. to 300 lbs. to the acre, would perhaps be better still. It will pay if we get $1 50 per bushel for wheat. At $1 per bushel the profits from the use of guano will be very slight, and may be on the wrong side of the ledger.

Gypsum, or sulphate of lime, seldom does any good on wheat in western New York, although it has a very good effect on clover, and sometimes on peas. Some good farmers sow a bushel of plaster (gypsum) on the wheat in the spring, but it is done, not to benefit the wheat, but for its effect on the clover sown with the wheat.

In regard to the time of sowing wheat, we have to steer between the Hessian fly and the midge—the *Scylla* and *Charybdis* of the wheat-grower. If we sow too early, there is increased danger from the Hessian fly, which deposits its eggs in the young plants in the fall; and if we sow late, the probability is that the midge, which deposits its eggs in the grain when in bloom, will destroy it. In western New York, from 1st to the 10th of September is now considered the safest time. As we go south, the wheat is sown later, but ripens earlier, and I believe we should find it to our advantage to get seed wheat from a southern rather than a northern latitude; but there is some difference of opinion on this point. It seems probable, to say the least, that the wheat would, for a year or two, retain a *tendency* to ripen at the same time it did at the south. The importance of this question will be seen when it is understood that if we could get wheat in bloom 10 days earlier it would receive little injury from the midge, and if it could be sown later, as at the south, the Hessian fly could do it no harm.

We have an early wheat—the Mediterranean—which generally escapes the midge, but it is of comparatively poor quality, though it improves much in this respect by cultivation. In regard to the quantity of seed per acre, I am in favor of rather thick sowing, say two bushels and a peck per acre if sown broadcast, or two bushels if sown with the drill. If the land is in fine tilth and high condition, less seed will be required. I know the quantity I have recommended is unusually large for this country; I know that a much less quantity is amply sufficient to seed an acre if the seed all germinates and the plants are not winter-killed; but we must sow enough to guard against these and other casualties, and I think I am warranted in saying, that thick seeding has a tendency to produce early wheat. This at least is certain: where wheat is thin from having been partially killed by snow-drifts or by what is known as "winter kill," the crop is always late, and generally suffers from midge and mildew. It is true that this late ripening may be owing to the same causes which produced the destruction of the plants. I know of no decisive experiment bearing on the point, but it is the opinion of several intelligent wheat-growers in western New York that thin seeding gives late crops. An experienced English writer contends that there is no advantage in drilling wheat unless it is hoed afterwards in the spring. This may be true of England, where the soil at the time of seeding is always moist enough to insure germination, but in this country, where we sow earlier and the soil is dry, there is this advantage in drilling: the seed can be deposited evenly, and at sufficient depth to insure germination. For this idea I am indebted to John Johnston; it cost me nothing, and I give it freely; but I believe he obtained it at a cost of five or six hundred bushels of wheat in one year.

On the cultivation of Indian corn my remarks shall be very brief. Corn will grow on all soils, from the lightest sand to the heaviest clay, among granite rocks and on the richest bottoms. It does not need so compact and calcareous a soil as wheat. It delights in a loose, friable, warm, porous, deep soil, abounding in organic matter. It does well on all good wheat soils, yet it often does better on soils too light and mucky for wheat. It is a gross feeder. We can easily make land too rich for wheat, but I have never yet seen any too rich for the production of Indian corn. Like all spring crops, corn requires an active soil. Its growth is very rapid. The atmosphere should have free access; fine tilth is essential; the soil should be made as fine as possible before planting, and after the plants are up the hoe and cultivator cannot be used too much during the first month. Throughout the vast corn-growing region of the west, if we can remove stagnant water, prepare the land properly, plant in good season, and use the horse-hoe freely, the soil is, in the majority of cases, rich enough to produce fair and remunerative crops. I have been in a two hundred acre field in Ohio, that has produced annually a good crop of corn for over fifty years without manure; but it was thoroughly cultivated. Not a weed or blade of grass was to be seen. In passing over the magnificent prairies in Illinois, I was much struck by the decided difference of the corn crops. Wherever the soil was dry, and proper care had been exercised in preparing the land, and keeping it well cultivated, the crops presented a most luxuriant appearance; but where careless preparation, and negligent, slovenly culture were rendered visible to the observant eye by the growth of weeds, the crop was as yellow and sickly as though it had got the ague. It was literally starved in the midst of plenty. Whether grown at the east or the west, on rich land or poor land, corn must have good culture, and I would here say that taking everything into consideration, as much energy and skill are necessary to produce *profitable* crops of corn at the west, as at the east. At all events, the difference is not as great as is generally supposed. Levi Bartlett states, that of thirty-five crops of Indian corn offered for premiums in Massachusetts, the average profit over all expense exceeded $51 per acre.

Corn will succeed on land that is too low and mucky for wheat, but though this is true, it is vain to hope for good crops if the land is surcharged with stagnant water. All the sunshine of our hottest summers cannot make such land warm. The heat is expended in evaporating the water instead of warming the soil. In passing along the various railroads of the country, I have been often saddened at the sight of thousands and tens of thousands of acres planted to corn, which by a little under-draining would have produced magnificent crops of this king of cereals, but which presented a miserable spectacle of yellow, sickly, stunted, half-starved plants, struggling for very life. Until the land is freed from stagnant water, all our efforts to produce good crops of corn will prove ineffectual. When this is accomplished, good cultivation will be most abundantly rewarded.

I have made some experiments with manures for Indian corn, on a field which had been under a scourging system of cropping with the cereals, and had never been manured for twenty years.

Unleached wood ashes had no effect on the corn, in this field; and 300 pounds of super-phosphate of lime per acre, though it gave the plants an early start, produced at harvest no larger a crop than 100 pounds of gypsum. But whenever ammonia was used, the crop was materially increased—more than *doubled* in one instance. The only deduction I would draw from this is, that the majority of our soils, relatively to ammonia, are not deficient in potash, soda, and phosphoric acid, so far as the growth of corn is concerned.

It is quite probable that there are soils where ashes and

phosphates may be needed for corn; but where such is the case, it is certain that they are much more needed for the growth of clover and other leguminous crops, and turnips, and that we cannot obtain from natural sources sufficient ammonia for the corn without growing these crops, or others which, like them, by their growth and consumption on the farm, furnish an increased quantity of ammonia for the use of the cereals.

FIFTEENTH DAY.—Feb. 17, 1860.

Mr. John Stanton Gould's lecture to-day was devoted to a classification and description of the grasses, with practical hints at the best varieties for farm use. After making some statements respecting the classification of the grasses, Mr. Gould proceeded to speak of the several species, describing their botanical and chemical characters, and the soils and localities to which they were severally adapted. With the grasses before him, he pointed out the marks by which timothy was identified and distinguished from others which resembled it. The largest stalk that he had ever seen was six feet six inches long, with a spike measuring eleven inches. The heaviest crop that he had ever heard of was on the farm of John Fisher, Carroll county, Md., who cut from an acre five tons, 1,622 pounds of dry hay. The heaviest crop of pure timothy that he had himself seen was on the farm of the Hon. Geo. Geddes, of Syracuse, which gave three tons to the acre. According to the analysis of Mr. Way, timothy yields more dry hay from a given amount of grass, and more of albuminous, fatty, and calorifacient matters from a given amount of dry hay, than any of the grasses upon which he experimented. But it must be remembered, that Mr. Way did not analyze either Poa compressa or Poa serotina.

The great drawbacks to its utility as a permanent meadow-grass are,—the very little after-math it produces; its liability to run out after two or three years; and the injury it receives from insects, with which it is infected, and which seem to be on the increase. The proper time for mowing timothy is just when the first dry spot appears above the first joint. If mowed before, the plant is injured. If left to a later period, the starch and sugar are converted into indigestible woody fibre, and the nitrogenous compounds, on which its value chiefly depends, are transferred from the leaves and culms to the seed, which mostly drop out before they reach the margin. Timothy is not well adapted to hot sands, gravels, and chalks, nor for hard, sterile clays; but thrives on peaty, damp soils, and especially on most calcareous loams, where it exhibits its fullest perfection.

Meadow Foxtails.—There are five varieties of the genus (Alopecurus), viz.: A. pratensis, A. agrostis, A. geniculatus, and A. aristulatus. The A. pratensis may be distinguished from its allied species by the equality of length in the glumes and paleæ, and by a twisted awn twice the length of the blossom. It rarely exceeds three feet in length, and does not usually yield over one ton to the acre. It is very watery in its composition;—100 pounds of the green grass gives only 19¾ pounds of dry hay, while an equal quantity of timothy gives 42¾ pounds. If one ton of green timothy be worth $5, the foxtail will be worth $2 07, if Mr. Way's analysis can be relied on. It is found abundantly in some of our best pasture; is one of the earliest to start in the spring, and the first to mature its seeds; its aftermath is exceedingly abundant, starting up immediately after mowing, and if the weather be showery will, in a week or ten days, give a fair bite to the cattle. It is not well adapted to alternate husbandry as it requires three or four years to bring a meadow to full perfection. It is very difficult to procure good seeds, as many heads are entirely destroyed by the insects. It is better adapted to pasture than to meadow, flourishes most luxuriantly on rich, moist, strong soil, the production from a clayey loam being three-fourths greater than from silicious soil.

Setaria glauca—Is good for nothing in meadows and pastures; it should be exterminated as soon as possible, which may be done by a thin coat of horse-manure applied in the fall.

Dactylis glomerata, or orchard-grass, sometimes grows five feet high, and has produced five tons, 1,859 lbs., an acre. One hundred pounds of it produces thirty pounds of dry hay; it contains nearly as much of fat and flesh-forming matters as timothy, but contains much less of heat-forming matters. If the latter is worth $5 a ton, orchard-grass will be worth $3 59. It flourishes well in shady places, and receives its trivial name from its adaptation to orchards. It affords a very large amount of after-math,—starts very early in the spring, and continues to send out leaves until late in the autumn. It shoots up very rapidly after mowing. Its disposition to grow in tussocks may be prevented by harrowing and rolling in the spring. It flourishes well on almost all soils and climates, but a sandy loam seems best adapted to bring out all its good qualities. On whatever soil it may be grown, the cattle will eat it in preference to any other, and will adhere to it as long as any of it is left.

Poa pratensis, a Kentucky blue-grass, in this section does not grow higher than 2½ feet, and cannot be relied upon to yield more than a ton and a half to the acre. One hundred pounds of the grass yields thirty-two pounds of dry hay to the acre, and is worth $3 20 per ton when timothy is worth $5. Butter made from this grass will keep sweet longer than that made from any other species. Its after-math is very luxuriant, and it stands the cold better than any other, but is liable to burn up in hot, dry weather. Its favorite locality is a limestone soil.

Poa compressa,—Wire, or blue-grass, has never been analyzed, but is believed to be the most nutritive of our grasses; it is certainly the heaviest, and grows about twenty inches high, standing thinly on the ground. It causes an abundant flow of very rich milk, and horses fed upon it alone will do as much work and keep in as good order as when fed upon timothy and oats combined. Sheep fatten astonishingly upon it, and all grazing animals eat it with avidity.

Agrestis vulgaris—Red-top, grows about 2½ feet long, and yields about 1½ tons to the acre. It is not a first-rate grass, but seems to be better relished by working oxen than by any other stock. It grows in very moist land.

Agrestes alba, or white-top, seems better adapted to sandy soils than the preceding, but resembles it very nearly in its botanical character.

Mr. Gould described many other varieties with much minuteness, illustrating their peculiarities from specimens in his hands.

The morning lecture by Mr. THEODORE S. GOLD of this State was on Root Crops—the field turnip, ruta-baga, beet, carrot, and parsnip—the soil they severally required, their culture, composition, and uses.

Root culture, says Mr. Gold, is the basis of successful English farming. As a means of supporting an increased stock, of supplying an abundance of enriching manure, and in thorough culture thus preparing for other crops, its value there proves inestimable; and there is no doubt that its more extended introduction here must be one of the means of securing that high degree of productiveness which constitutes the most successful agriculture. The estimated value of the root crop of Britain amounts to £20,000,000, or upward of $100,000,000, while its subsequent advantages, as preparatory for other crops, vastly exceed this sum. It was a remark of Daniel Webster that, "Take away turnip culture, and England would become bankrupt."

The turnip belongs to the same botanical genus as the cabbage, which also embraces in its varieties the cauliflower and broccoli. Two or three species are made by some botanists of the turnips, which exhibit such great variations in form and color, while others embrace them all in one. No class of plants exhibit greater adaptation to the various conditions to which it is subjected by culture, and though they have been long known, it is but recently that they have acquired any importance as farm crops. Hence we may anticipate a high degree of improvement in the future. While the average of the turnip crop of the State of New York is shown by Mr. Randall to be only 88 bushels per acre, this is far below the capacity of the soil as is proved by the reported premium crops, reaching, in one instance, as high as 2,102 bushels per acre. The details of management in the case of this crop were given, in the language of the cultivator, J. T. Andrew, of West Cornwall, Conn., to show what results may be attained by skilful culture. New land produces the best turnips for all purposes, especially for table use. Sow white turnips in drills, or broadcast, the latter part of July; ruta-bagas the last of June, in drills, twenty-five to thirty inches apart. Quantity of seed, one pound per acre. The most thorough preparation of the soil by deep and careful plowing, and early and repeated tillage by the horse- and hand-hoes, are necessary in the highest degree in this and all the other root-crops. The ruta-baga is a gross feeder, and requires an abundance of manure either in a raw state or fermented. This may be applied broadcast, or under the drills. Bones and super-phosphates are considered essentials to turnip culture in England. My experiments with them have proved quite undecisive as to their value here. Early thinning to a distance of twelve inches in the row is required for the largest produce. If sown late, for table use, they may stand much closer.

The *beet* in the form of the *sugar beet* in France and Germany, and the *mangold wurtzel* in Great Britain, is taking a position of more importance than even the turnip. It requires much the same culture as the ruta-baga, while the greater yield of the mangold, its freedom from disease and the attacks of insects, and its superior keeping qualities, render it a general favorite, while its fitness for enduring heat and drouth especially adapt it to our wants. The quantity of seed varies from two to four pounds, according to the manner of sowing. The drill sows it very unequally, from the rough surface and varying size of the seed capsules. It is better sown by dibbling with some instrument, at regular distances of twelve inches in the drill. Sow in May or June, about the time of planting corn, and harvest before severe frost. It keeps admirably, even till the new crop grows again. It is not considered fit for use in England till after Christmas. It is excellent for sheep, cattle, and swine. The latter prefer it to potatoes or carrots. Twenty pounds is not a very large size for this root. The lecturer here exhibited one of his own raising, weighing 20 lbs. The amount per acre of 1,200 or 1,500 bushels is here considered a very good crop, while in France and Germany reports are given of crops almost exceeding belief. Mons. Auguste de Gasparin, in the *Journal d'Agriculture Pratique*, reports having raised on one-fourth of an acre 127 tons of 2,000 pounds each, or 5,080 bushels of beets, at 50 pounds per bushel. He also states that Mons. Koechlin, in Alsatia, raised at the rate of 156 tons per acre, or 6,240 bushels. The roots averaged 37½ lbs. each, and as this allows five square feet for each plant, it is quite within the limits of possibility.

The carrot is the most esteemed of all the roots for its feeding qualities. When analyzed it gives but little more solid matter than the other roots, 85 per cent. being water; but its influence in the stomach upon the other articles of food is most favorable, conducing to their most perfect digestion and assimilation. This result, long known to practical men, is explained by chemists as resulting from the presence of a substance called pectine, which operates to coagulate or gelatinize vegetable solutions, and favors this digestion. Horses are especially benefited by the use of carrots. In that true "high farming" which is most eminently profitable, the culture of roots holds an important place. It requires labor and requires capital; but the foolish system of *labor-saving*, by abstaining from its use, lies at the foundation of very much of the wretched farming with which we are so justly charged. In that happy condition of Connecticut agriculture in which every acre in this State shall either support its cow or produce its equivalent in value for animal or human food, successful root culture must exercise an important part.

SIXTEENTH DAY.—Feb. 18, 1860.

Hear what old Mr. Levi Bartlett, of New Hampshire, said yesterday in opening his farmer-like lecture on the cultivation of winter wheat in New England: "It may be asked why one so conscious of oratorical defects, should attempt speaking at all, especially in such a convocation as this. I can only answer in the words of the wily old Roman, that I am a plain, blunt man, who loves the cause; and therefore am I come to speak, but most of all to hear, in this assembly. And if forty years of study of the principles of agriculture, and full twenty devoted to practice, with an enthusiasm which time has not abated, give me any claim on your attention, then I trust to your generosity to excuse the manner for the sake of the matter." Considering that the matter was of an eminently practical character, and that friend Bartlett's quaint jokes kept the convention in a roar, his apology was scarcely needed.

Mr. Bartlett said that from his earliest recollection down to 1852, spring wheat was the only kind raised in New Hampshire. In fact, he never saw a field of winter wheat until he was fifty years of age. Spring wheat had, in general, been pretty successfully grown on all land that would produce corn, until the appearance of the midge, some quarter of a century ago. The ravages of this pernicious insect were so great, especially on valley farms, that the culture of wheat was in great part abandoned, so that a large part of our farmers, as well as those of all other professions, depended upon Western and Southern flour for their wheaten bread; and as there was but little else eaten, it has been a mystery among our most acute financiers how the people paid for all this "boughten flour." But within the past six or eight years, matters in this respect have greatly mended, in consequence of many of our New Hampshire farmers having turned their attention to the culture of winter wheat, in which most of them have been very successful.

In the summer of 1852, the son of his (Mr. Bartlett's) neighbor was in western New York, and was so pleased with the fields of winter wheat, that he took home with him fourteen quarts of the "bald" variety of white wheat grown there. This was sown on about one-third of an acre of dry, loamy land. From a combination of favorable circumstances, it yielded sixteen bushels of prime wheat, at the rate of forty-eight bushels per acre. Nearly all of the sixteen bushels was readily sold for seed at $3 per bushel, and as was to be expected, under the excitement and the entire ignorance of its proper culture by the farmers, some succeeded well, while others made a partial or total failure. In 1853, he sowed one bushel on light, pine land, from which a crop of beans had been removed at the time of sowing the wheat, he applying to the land one hundred and fifty pounds of Peruvian guano. The wheat was sown 20th of September, at least twenty-five days too late. The yield was about nine bushels. For the five past years, he has been experimenting with winter wheat on a variety of soils, and with different manures. He has grown it on intervale lands, on hills, on light, dry soils, and stiff, heavy ones. These last, however, have always been ridged up, turnpike like, and the dead-furrows well cleaned out to drain off the water. Sometimes the wheat has been sown on a newly inverted timothy sod; at other times on a clover ley, and upon wheat and oat stubble. In every instance the land has been pretty liberally manured with farmyard manure, or guano. During the six or seven years he has grown it, it has suffered but very little from winter-killing, nor has it been injured to any great amount by the midge, although his own spring wheat, and that of his neighbors, has been nearly ruined by that insect. As an illustration of this, he stated that, in 1857 he harvested twenty-eight bushels of prime winter wheat from seven pecks sowing of the previous autumn. From a bushel of spring wheat, sown in May, 1857, he harvested but seven pecks, and of a very poor quality at that. His crops have averaged about fifteen bushels to the bushel of seed sown; many of the farmers in his vicinity have raised twenty bushels, and over, from the bushel of seed sown; and one farmer raised on "hill-land" last season, twenty-two bushels from a bushel of seed, while another, on a low-lying farm, grew ninety-one bushels from four and a half bushels of seed. These "out west" might not be considered very great crops, but they are more than twice as large as those of spring wheat, in his section of New Hampshire, have averaged of late years.

He has been experimenting for several years with a great variety of Patent Office wheats. Out of the number only four varieties have been found adapted to his place. Of these the "early Japan," the original of which was brought from Japan by the late Commodore Perry, is a red wheat, some ten days earlier than any other variety he has grown, its earliness putting it beyond injury from the midge. The "Tuscan wheat," from Michigan, which was distributed by the Patent Office, was accompanied by a certificate from several Michigan farmers, which showed that it had been grown there for seven years, and had never been known to rust. It is a large-grained, flinty variety, yielding fifty pounds A No. 1 flour to the bushel. The "Early Noè," the original seed of which was procured from France, has the merit of early maturity, as it was said to be ten days earlier than any other grown in the dominions of Napoleon. With Mr. B. it has not proved earlier than his other varieties. It has a good-sized kernel, and very stiff, white straw, and promises to be a variety worthy of general cultivation. General Harmon's "improved white flint," from the Patent Office, is a most beautiful wheat; hardy, productive, and making the finest quality of flour and bread. Also, another variety of white wheat, yielding fifty-seven pounds of fine flour to the bushel.

Samples of all the above varieties, both in the straw, and the grain in bottles, were exhibited during his lecture, which fully sustained his positions in regard to the adaptation of our New England soil and climate to the profitable production of winter wheat. He usually carries four bushels of his wheat to mill, to make a barrel of flour, and pays for the grinding some thirty cents; and he finds a material difference between this, and handing over a ten dollar bill, or giving his note for that amount for a barrel of Milwaukie or Chicago flour.

To insure success in raising winter wheat in New Hampshire, the land must be dry, in good heart, and well-worked. The seed should be sown from the 20th of August to the 5th of September. It should be thus early sown to have it get well-rooted before winter, and to hasten its maturity, so as to escape the midge. A difference of five or ten days in the blossoming of a field of wheat frequently makes the difference between a very good, and a very poor crop. This is owing to the midge. He has, by sowing early, escaped loss from the midge and rust, while some of his neighbors, who have delayed sowing till after their corn was harvested, have suffered by winter-kill, midge, and rust.

Learning that Col. Cate, of Northfield, N. H., had been very successful in growing winter wheat for a number of years, Mr. B. wrote to him upon the subject in December last. He read an extract from the Col.'s letter, which is as follows:

"I commenced the cultivation of winter wheat in the year 1850, and have continued it without interruption up to the present time. The first year I sowed one bushel of the 'white-bald winter wheat,' on the 6th day of September of that year, on land which had grown a crop of corn the same season. The land had been tolerably well manured in the spring; but from some cause, I hardly know what, did not produce a large crop of corn. The wheat came up well, and tillered finely during the autumn following. When winter set in, it stood all over the piece ankle-deep, and quite thick. In the spring following, and before the warm weather set in, it seemed to retain all its freshness of color and vitality. It did not suffer in the least from the winter cold, nor the spring frosts. It was harvested in July, and by my record of crops I find it was threshed August 7, 1851. It measured up, of clean wheat, twenty-four bushels, and weighed sixty-five and a half pounds per bushel. As already said, I have continued to raise winter wheat ever since, and am perfectly satisfied that it is safer, by far, and surer than summer wheat, for most soils in our State.

"My method of culture has been briefly as follows: In the first place, I have cultivated on ground which had been hoed, and on the inverted sod, breaking at or about the time of sowing. Out of the time I think I have sowed four years on the recently broken up land, and I do not see but that I have succeeded in one case as well as in the other. I hardly need say that the land in either case should be thoroughly plowed and harrowed. I have invariably soaked my seed in a strong solution of salt and water, and most of the time have used 'Glauber's salts' with the common coarse salt—not, however, soaking the seed more than two hours. After draining it, I have generally rolled it in ashes, and then sowed immediately. If my land has been cultivated and manured the spring before, I use no other manure or stimulant at the time of sowing. If not, as in the case of newly broken up land, I have used, and am so well satisfied with the results that I shall continue to use, from ten to fifteen bushels of ashes, with from one to two bushels of salt, per acre, sown broadcast over the field at the time of sowing the seed. The result has always been a larger crop than under the most favorable seasons I could get from spring wheat sown on the same kind of soil, and side by side."

Col. Cate, as well as Mr. B., thinks that winter wheat can be grown with as much certainty in New England, as it is by western farmers, but, not as cheaply—for here we must use manure to obtain good crops.

Mr. GOULD'S third lecture, to-day, was devoted to a description of the grasses and clovers, in continuation of his lecture yesterday. He denied the distinctions of the genus *Festuca*, as laid down in botanical works, asserting that F. ovina and F. rubra were merely variations of F. duriuscula, and that F. loliacea and F. pratense were varieties of F. elatior. It is sufficient for all the purposes of the farmer to divide the genus into two classes :

1. Those having more or less hairs on the leaves ; and
2. Those having smooth leaves.

This genus affords us some species that are of great value in an agricultural point of view, each of which, under certain circumstances, is of great value, and very permanent in its forms and qualities. Thus: *F. ovina* is essentially a grass of the thin soils resting upon rocky uplands, as on the mountain limestone and most mountain ranges.

F. duriuscula.—In the valleys between such hills, and in the more sheltered pastures of the upland districts.

F. rubra.—In the more sandy loams of the lowland meadow, and by the sea-shore.

F. loliacea.—Rich meadows on river banks, or under irrigation.

F. pratensis.—Best lowland meadows, not liable to floods.

F. elatior.—On sandy clays, or other stiff and strong lands, especially on the sea-shore.

The festucas are invariably present in our best pastures, and especially present in those of the most famous cheese districts.

The *F. pratensis* is worth $3 33, where timothy is worth $5, per ton. It follows next after meadow fox-tail as an early grass, and affords a bite earlier than orchard-grass.

He gave the *Bromus* family a very bad name, adducing a number of experiments to show that it was neither agreeable nor nutritious to cattle. *Bromus erectus* was said to be the only perennial species in the genus. Early mowing was recommended as a means of extirpating this family. Pheasants are exceedingly fond of the seeds, and frequently pick off the spikelets before the seeds are ripe, that they may enjoy the much coveted luxury.

Lolium perrenne, or Rye-grass, is still the favorite grass of England. It occupies there the same place that timothy does with us, and is probably better adapted to a wet climate like England than to a dry one like ours. Sixty varieties are cultivated in England of this one species. One of the most remarkable of these is the viviparous Rye-grass, which grows there with great luxuriance. After midsummer it is strictly viviparous, never producing either flowers or seeds, but young plants from the glumes, which, when the original plant is supported, will produce new plants from two to three inches in length.

Lolium Italicum, Italian Rye-grass, is worth $2 69 when timothy is worth $5. One hundred pounds of it give twenty-four and a half pounds of dry hay. It is best adapted to limestone and light soils, and is one of the most desirable varieties for irrigated meadows.

Triticum repens, known as "quack," "twitch," or "dog" grass, is very easily recognized by its spikelet of eight- or ten-awned flowers placed *flatwise* toward the sachis. It is a terrible pest in alternate husbandry, growing in all sorts of soils, and robbing the cultivated plants of the richest portion of their food. In very dry seasons it may be killed by plowing it very thoroughly in July, and sowing the ground with buckwheat. Its culms (stalks) sometimes attain an altitude of three feet, but it ordinarily stands two feet high. It forms a tolerably good hay, and is much relished by the stock as a pasture grass. It operates as an emetic on dogs; and is very useful in binding the sloping banks of railroads.

Anthoxanthum odoratum, Sweet-scented vernal grass, is not very valuable for hay or for pasture, as one hundred pounds of it give only nineteen and three-quarters pounds of dry hay. An acre only yields three-quarters of a ton of dry hay. It starts very early in the spring, and continues to throw out leaves during the summer. Its after-math is more valuable than the first growth, and is supposed to communicate the peculiar flavor which characterizes the Philadelphia butter.

Glyceria nervata grows in wet places. Its culms (stalks) are extremely succulent; it is the hardiest grass in existence, and always grows more vigorously after a severe winter than after a mild one.

Poa serotina, or Fowl-meadow, is one of the earliest grasses cultivated in this country, and is still among the best. It does not injure by standing, as do other grasses; but may be cut at almost any time. Hares and rabbits are extremely fond of it. It is easily made into hay, and never seems hard or harsh, and produces sound seeds in great abundance.

Trisetum subspicatum is a mean, stingy grass, growing on stiff, clayey hill-sides which have a northern aspect. It is only fit to be grown on soils that will bear nothing else.

Zizania aquatica.—Mr. Gould spoke of this grass as growing in places that were wholly covered with water. It is very sweet and nutritious, and cows fed upon it have a copious flow of milk. In favorable situations it produces five or six tons to the acre, growing to the height of nine feet. Many birds, especially the rail, fatten on it in autumn. The Indians collected its seeds, which resemble rice, and stored them for winter use.

Mr. Gould spoke at great length of the clovers, detailing many interesting facts in relation to them, and giving much practical advice respecting their cultivation. He especially recommended the increased culture of lucerne (*medicago sativa*). The best soil for it is a sandy one, resting on a porous calcareous subsoil. Its roots penetrate fourteen feet in depth, and hence a hard subsoil is fatal to successful growth. It arrives at its greatest perfection after three years. In one recorded case, eleven acres sufficed to keep eleven horses two hundred and ninety-nine days. In another, a field of eight acres kept eight horses three hundred and fifteen days. In both cases a large number of sheep were fed on the ground after the last cutting for the horses. Chancellor Livingston, in Columbia county, N. Y., cut twenty-five tons from an acre in five mowings. It is ready for cutting about the first of May, and may be cut over every thirty days thereafter. It is remarkably adapted for milch cows, where the milk is sold in the market, but butter made from it is not so sweet as from other grasses. It is greatly relished by both horses and cattle; one hundred pounds of it will make twenty-five pounds of dry hay, and its nutritive powers bear such a relation to those of timothy, that it is worth $3 13 per ton, when that grass is worth $5.

The only difficulty with lucerne is, to get it started. It must be sown in drills, and carefully hoed until it is large enough to cover the ground. If this precaution is taken, and a drouth does not occur just as the young plants are starting, it will be pretty sure to succeed, and *will last for twenty-five or thirty years.* If, however, it is overrun with weeds in the beginning, or a severe drouth occurs, it grows feebly and soon dries out. The seed is covered with a very hard and compact coat, which, if the weather be dry, will greatly retard vegetation. It is, therefore, generally the practice to steep it in warm water, to soften the coat, for six or eight hours before sowing. From fourteen to eighteen pounds of seeds are usually sown on an acre; but, as many of the seeds are imperfect, and as fine and succulent plants are more desirable than coarse and rank ones, it is better economy to sow twenty-five pounds.

The following table gives the comparative value of lucerne and common pasture. After being kept on lucerne for about ten days, the milk of three cows was separately measured, and the produce in Scotch pints was, on the 28th of May, as follows:

No. 1.—Calved in March, gave........................ 13 pints.
No. 2.—Calved in January, gave...................... 10¾ "
No. 3.—Calved in May, gave.......................... 10 "

They were then put alternately in pasture and lucerne during the following periods, when the produce was found to be:

	Pasture.	*Lucerne.*	*Pasture.*	*Lucerne.*
	To June 8	To June 13.	To July 13.	To July 19.
No. 1.	—12½ pints.	12¾ pints.	10 pints.	11 pints.
No. 2.	— 9¼ "	10¾ "	9¼ "	10 "
No. 3.	—10½ "	10 "	9 "	8¾ "

Mr. Gould spoke at length of sainfoin, tares, and succory, and after the conclusion of the lecture he exhibited the various grasses of which he had spoken, to the more zealous students, and gave them particular instruction in the botanical analysis of the different genera and species. He urged them very earnestly to make themselves experts in the botany of the grasses, assuring them that this was essential to the acquisition of a correct knowledge respecting them. And I am happy to know that a large number of the students expressed themselves determined to enter vigorously on the study of the grasses, and forage plants of our country.

SEVENTEENTH DAY.—Feb. 20, 1860.

This, the fourth and last week of the course, is especially devoted to the subject of stock-breeding; but Professor B. Silliman, Jr., gave us this morning a lecture on Meteorology, devoting the hour to a very simple and elementary discussion of the phenomena of the atmosphere as respects the fall of rain and the distribution of temperature, describing the thermometer, hygrometer, and rain-gauge.

He spoke briefly of climates, and seasons, and the influence of the sun, not only in causing the differences of seasons, but on the mean daily temperature. The mean daily temperature at Philadelphia had been found to be one degree above the temperature at 9 A. M. The average annual temperature of the atmosphere diminishes from the equator towards the poles.

But the temperature is not the same for places in the same latitude in the two hemispheres, as is seen in the following table:

PLACES.	LAT.	TEMP.	PLACES.	LAT.	TEMP.
Falkland Isles,	51° S	47° 23	London,	51° 31′ N	50° ·72
Buenos Ayres,	34° 36′ S	62° 6	Savannah,	32° 05′ N	64° ·58
Rio Janeiro,	22° 56′ S	73° 96	Calcutta,	22° 35′ N	78° ·44

This variation is owing to a variety of local causes, such as the elevation and form of the land, proximity to large bodies of water, the general direction of winds, etc.

The temperature of the air diminishes with the altitude. As a general rule, it may be stated that there is a diminution in temperature of 1° F. for every 343 feet of elevation. On rising from near the level of the sea, the rate of decrease is more rapid; after a certain height is reached it proceeds more slowly; but in very elevated regions it again increases.

It follows from this that in every latitude, at a certain elevation, there must be a point where moisture once frozen must ever remain congealed. The lowest point at which this is attained is called the limit of perpetual snow, or the snow-line. This point is highest near the equator, and sinks towards either pole, as is shown in the table.

PLACES.	LATITUDE.	SNOW LINES.
Straits of Magellan,	54° S	3,760 feet.
Chili,	41° S	6,009 "
Quito,	00°	15,807 "
Mexico,	19° N	14,763 "
Ætna,	37° 30′ N	9,531 "
Kamtschatka,	56° 40′ N	5,248 "

Isothermal lines were very briefly illustrated from a map of the United States, on which were traced from the map in the Patent Office Report for 1856–7, the lines of summer and winter temperature in various latitudes. The great value and importance of such researches to agriculture were insisted on by the lecturer as giving the only rational explanation to anomalies of climate, etc., otherwise inexplicable. The great contrast between the latitudes and isothermes of wheat and other grains, of the limits of the vine, of maize, etc., was pointed out on a chart, and in this connection the summer climate of British Columbia was alluded to. He also called attention to the marked difference in the winter climates of the two oceanic borders of the continent, as compared with the corresponding latitudes in the interior.

The Aqueous Phenomena of the atmosphere were next considered. The presence of moisture in the air at all times was explained, its amount depending on the temperature.

That the capacity for moisture is greater as the temperature increases was shown by the following table.

A body of air can absorb:

At 32°	F.	the	160th	part	of its	own weight	of watery vapor.
" 59°	"	"	80th	"	"	"	"
" 86°	"	"	40th	"	"	"	"
" 113°	"	"	20th	"	"	"	"

It will be noticed that for every 27° of temperature above 32°, the capacity of air for moisture is doubled. From this it follows, that while the temperature of the air advances in an arithmetical series, its capacity for moisture is accelerated in a geometrical series.

The lecturer here exhibited various forms of *hygrometers*, and illustrated their use experimentally:—Saussure's hair hygrometer, various *hygroscopes*, Daniells' condensation hygrometer, and August's hygrometer of evaporation. He also exhibited a simple substitute for the costly condensation hygrometer, being nothing but a bright silver goblet or tumbler containing water and lumps of ice. The first condensation of dew on the polished metallic surface is watched for, and the instant it appears the difference between the thermometer in the iced water and the air is noted. This gives the dew point, or temperature at which fog would be produced.

The mode of measuring the rain fall was also described. One of the simplest rain-gauges was a cylindrical vessel of tin, or copper, furnished with a float: the rain falling into the vessel, the float rises. The stem is graduated so that a depth of water of one one-hundredth of an inch is easily measured.

The unequal distribution of rain over the surface of the earth was touched upon, and the influence of mountain ranges was pointed out in causing precipitation of rain. As a general rule the amount of rain was in proportion to the average temperature; or, what is the same thing, to the amount of evaporation: local causes, however, very greatly modify this general rule. The number of rainy days bears no proportion (or an inverse one) to the amount of rain which falls in particular latitudes. Thus while the yearly fall of rain in the tropics is ninety-five inches, there are not over seventy rainy days; while here, with an annual rain fall of about forty inches, we have one hundred and thirty or more rainy days. The following table shows that the ordinary rains of the tropical regions are more powerful than those of the temperate regions.

M. LATITUDE.	MEAN ANNUAL NUMBER OF RAINY DAYS.
From 12° to 43°	78.
" 43° " 46°	103.
" 46° " 50°	134.
" 50° " 60°	161.

In the northern part of the United States there are, on the average, about 134 rainy days in the year; in the southern part, about 103.

The greatest annual depth of rain occurs at San Luis, Maranham, 280 inches; the next in order are Vera Cruz, 278; Grenada, 126; Cape François, 120; Calcutta, 81; Rome, 39; London, 25; Uttenberg, 12·5. In our country the average annual fall is 39·23 inches; at Hanover, N. H., 38; New York state, 36; Ohio, 42; Missouri, 38·265.

Prof. Silliman illustrated these general principles by an analysis of the average results observed by Dr. S. P. Hildreth, at Marietta, Ohio, Lat. 39° 25′ N, and Long. 4° 28′ W of Washington city, for 31 years, from 1828 to 1859. It appeared from these tables that the rain fall at Marietta varied from 61·84 inches in 1858 (the wettest year in 40, and one in which there were only 170 fair days) to 32·46 inches in 1856; the average of the whole period being 42 inches.

The time permitted only a cursory allusion to the other aqueous phenomena of dew, frost, and hail. The lecturer pointed out the defects of common thermometers, and the mode of selecting a good one. He remarked that between 32° and 212° Fahrenheit it was easy to select an instrument which would indicate the temperature within one or two degrees of accuracy. He exhibited, however, four instruments taken that day from the stock of a dealer, from which he read as follows: 64°; 62°; 65°; and 66°. Below 32° common thermometers were generally very unreliable; the difference amounting near zero often to more than 10°. He stated that in old thermometers the point of freezing (32°) was found almost uniformly too high, and that the readings of old thermometers were as a rule too high. This was owing to a permanent displacement of the zero point, partly arising from atmospheric pressure on the surface of the ball, and partly from the slow contraction of the glass subsequent to the heating to which it was subject in filling.

He gave practical rules for the exposure and observation of thermometers. A thermometer should never be hung against the wall of a house, for the radiated heat makes the mercury rise often as much as 4°. It should be placed on a post in the yard. It has been proved that in our country the temperature at 9 A. M. will be just 1° less than the average of the whole day. If our thermometer marks 50° at that hour, we may know that the day will average just 51°. The coldest hour of the day is 7 A. M., and the warmest 2 P. M.

He concluded by commending to farmers the study of meteorology, as an important element of the practical education on which success in agriculture must depend.

Mr. Sandford Howard, of *The Boston Cultivator*, gave a

lecture on horses, at 3 o'clock. He referred to the great diversity of appearances between the heavy Flemish or English dray-horse, which will weigh a ton, and the little ponies that scamper over the hills of Shetland. The heavy horse will always be found in plain countries, and good and fertile districts. Horses may be divided into three classes:—first, gallopers, or runners; second, trotters; third, walkers. The lordly Arab steed of the desert is the type of the former class, as also is the so-called thoroughbred racer; the trim-built Morgan, of the second; and the heavy Conestoga and Clydesdale, of the third. The horse is not a native of America, but has been introduced at various points from various sources. The wild horses of Mexico and some South American countries have sprung from the animals brought over by the Spaniards. The German settlers of Pennsylvania introduced the heavy draught-horse of their fatherland. The French settlers of Canada brought another breed—the ancestors of the Canadian horse of to-day. The modern Norman, or Percheron horse, has been introduced into New Jersey. The English and Scotch of Canada West have brought over their Clydesdales and other draught horses. The race-horse has found a home in many parts of our country; and so all sections have derived their horse stock from the Old World.

For long distances, with a heavy weight on the back, at a galloping pace, the true Arab is the best model. For short distances, at headlong speed, and with light weights to carry, the English racer, or "thoroughbred," is required. Of trotters, for quick driving in light vehicles, the "roadster" best meets the requirements,—the best American horses of this description being probably superior to any in the world—certainly superior to the English. For city coach-horses, less speed and hardiness being needed, an animal of more size is called for; a purpose for which the Cleveland Bay, or a mixture of the race-horse with some large-sized stock answers well. For omnibuses and horse-railroad cars, a more muscular horse, able to endure hardship, is preferable; and the French "Percheron" is well adapted to this work. Of horses, the uses of which only require a walk, and where heavy burdens are to be drawn, a conformation more adapted to strength and less for speed is necessary. For heavy draught, some of the English and Scottish breeds are best. For farming work, where horses are wholly used, and for drays, carts, &c., of cities, the Suffolk and Clydesdale breeds would be preferable to the horses now generally used for these purposes in this country.

In general, and especially for racers, roadsters, and draught-horses, it is better to keep the varieties distinct, breeding each in reference to a standard or ideal. If experiments in crossing are made, they should be conducted with caution, and in such a manner as not to hazard a loss of the valuable properties already possessed by an established breed.

EIGHTEENTH DAY.—Feb. 21, 1860.

Mr. Charles L. Flint, Secretary of the Massachusetts State Board of Agriculture, and author of standard works on "Grasses and Forage Plants," and "Milch Cows and Dairy Farming," gave, to-day, in his first discourse, a number of valuable hints to dairymen, and much information of general interest. His lecture was listened to with great attention.

Mr. Flint called attention to the fact that the dairy qualities of our stock are artificial, and mainly the result of care and breeding. The cow, in her wild state, gives only enough milk to nourish her offspring for a short period, and then goes dry the rest of the year. The prime object of the farmer is to develop and improve her milking qualities, and hence he should select his cows with reference to the amount of food he has for them. Large animals require rich and luxuriant pastures, or they lose their fair proportions and deteriorate on a stinted nourishment. The objects of the dairyman should be kept in view in selecting his cows. The animal most profitable for a milk dairy may be very unprofitable for a butter or cheese dairy. The first cattle imported into New England arrived at Plymouth in 1624, and they are described as of a variety of colors. These, with the importations of Capt. John Mason, from Denmark into New Hampshire, in 1631–4, laid the foundation of the native stock of New England; and this stock must be regarded as an exceedingly valuable foundation for improvement, which may be effected either by careful and judicious selections, or by crossing with foreign and already highly improved breeds.

Grades are often more valuable for practical purposes on the farm than pure breeds. In breeding it is important to have a specific object in view, as for beef, milk, or labor—the complete union of these qualities being, to a considerable extent, impracticable. Great milkers are rarely very handsome animals. They seldom have the well-rounded forms of fattening animals, but are often coarser looking and more angular. In breeding to produce large milkers, it is especially important to select males that come from great milking cows—since the dairy qualities are transmitted more surely through the male offspring. The most celebrated dairy breeds are the Swiss, the Dutch, the Jersey, and the Ayrshire. The Jerseys give the richest milk, and the Ayrshires the largest quantity, in proportion to the food consumed and their size, and are very valuable as a means of improving our common or grade stock. But, whatever breed is selected, success will mainly depend on the care and management, and especially on the food. VERY LITTLE MILK COMES OUT OF THE BAG THAT IS NOT FIRST PUT INTO THE THROAT. It is poor economy to overstock the farm, as is too often the case: the cows come out of the stall in spring in no condition for the profitable production of milk. The cow should be regarded as an instrument of transformation; a machine for the manufacture of milk. The food is the raw material, milk the product —salable, and always in demand. The machine is the capital invested, costing nearly as much when not running as when running on full steam. How absurd, therefore, how unbusiness-like, for the farmer to slacken up the supply of raw material, or by neglect, exposure, or otherwise, to suffer the machine to get out of order, or to yield a product far below its natural capacity.

Regularity of feeding is next in importance to a full supply of nutritious food, and cows thrive better on a good and regular system, than on a larger amount fed at irregular intervals.

Cows in milk ought not to be exposed to cold in winter. They require less food and give more milk if kept housed. They ought not to be even turned out to water in extreme cold days, and they will be sure to fall off in milk if they are. The loss from a neglect of this precaution is often far greater than farmers are aware of. The cow should be kept in a sound and healthy condition by judicious feeding and exercise, but exposure in extreme cold weather is never advisable. Moist and succulent food increases the quantity of milk; dry food, as hay, alone, makes a thicker quality. Food rich in starch, gum, sugar, &c., increases the butter in milk.

Quietness also promotes the secretion of fat, and increases the richness of milk. Green grass is more nutritious and more digestible than hay, which, like all other coarse and dry food, is made more nutritious by cutting and moistening, or steaming. All ruminating animals require more or less bulky food, the bulk contributing to the healthy activity of the digestive organs. The most valuable additions to this branch of farming have been made by the elaborate and successful experiments of Mr. Horsfall, who found that he could make as much and as rich butter in winter as in summer. His whole course of management has been republished in this country in the appendix to the lecturer's Treatise on Milch Cows and Dairy Farming.

Particular attention was called to the management of young heifers, and the time when they should be allowed to come in, as well as to the care which should be taken to prevent any faulty habit or constitutional defect to become fixed upon them.

Suppose, for instance, a heifer should come in in winter, or in very cold weather, which would prevent the distension of the tissues of the skin, and she should be fed *on dry food*, which had little tendency to develop the milk vessels, or the organs of secretion. These organs will adapt themselves to supply a small yield of milk, and thus a habit may be fixed upon the animal for life, or which it might be difficult to overcome entirely afterward. Hence, some of the external signs of a good milking cow are found on animals whose product does not justify expectations.

A young cow with her first and second calf should be made, by judicious feeding, to give a large quantity, and to hold out well, and by gentle treatment, to be docile and obedient.

A certain shepherd-lecturer at a farm-school in Saxony, illustrates his lectures on breeding by presenting before his class sheep of various breeds and diverse qualities. So far as my information extends, it has never been attempted in this country before to-day,—when Mr. THEODORE S. GOLD placed on the stage a Cotswold, a Merino, and a Southdown. The latter arrived a little after the lecturer had concluded, but was seen by many then present. It is a new, and a most capital idea; and hereafter, he who will lecture on sheep without the living illustrations ready for reference, will be behind the age.

The sheep, as Mr. G. justly remarked, has been associated with man from the time of Abel, and in some countries is now the chief national wealth. In Saxony, not larger than Connecticut and Rhode Island, there are 3,500,000 sheep; England and Wales produce 26,000,000; while in the whole territory of the United States we raise only 21,000,000. It must be remembered that in the great sheep countries of Europe, farming has perhaps arrived at its greatest perfection of development—a circumstance which should weigh well with our farmers, whose poor hilly lands will barely keep them and their families above starvation, under the present cropping with Indian corn and the cereals.

The "felting" property of wool is due to the peculiarly rough or barbed character of its outside, which causes it to adhere together in mass, and a woollen garment to shrink and become thicker when washed. "Fulling" is another name for the same quality; "fulled cloth" being the name given to the article made by subjecting woollen cloth to the action of water, and pressure in a machine. By aid of the microscope, we see that the fibre of wool is covered with a multitude of leaf-like serrations (saw-tooth projections), pointing upward like the leaves on a shoot. The curved form of the wool fibre favors its felting, but it is to these million invisible hooks that we must look for an explanation of the property. Now, in the finer grades of wool there is the greatest number of these tentatious hooks in a given length, and hence their superiority for close textured and fine goods. This little explanation will give our farmer friends an insight into the subject of breeding sheep for various purposes. The Merino is, above all, the wool-maker of fine quality. Leicester wool is famous in England for combing, or worsted making, but is much coarser than Merino. "Yolk," or "gum," is the name of a glutinous secretion from the skin of the sheep, which coats and adheres to the wool. It is a true potash soap, and if it were not for the presence of free animal oil with which it is mixed, wool might be washed without the use of soap. It is most abundant in fine-woolled sheep, and is more largely secreted in the fat sheep than in a lean one.

It is very desirable to grow sheep that will have an equal degree of fineness of wool over a large portion of the body, and success in this respect marks the good breeder. "Trueness" is a term used to indicate the evenness of fibre in size throughout its whole length. When the sheep, from disease or want of food, becomes poor, the wool fibre is rendered weak and almost ceases to grow. When it starts again, it breaks easily at this weak point, being what is termed "breachy," and the wool is called "unsound." Its value is greatly depreciated by this circumstance. Let those who starve their sheep take the hint. The best health is obtained by neither over-feeding nor starving.

The lecturer gave sketches of the various outlandish breeds of sheep found in various parts of the world, among which were the "fat-tailed" family, the "fat-rumped" sheep of Asia, the many-horned sheep of Cyprus and Iceland, the Siberian, Tartarian, Russian, and others. It is not known if the Merino is a native of Spain. Beside that breed, there is in Spain another—a coarse-woolled, large variety, to improve which a number of Cotswold bucks were imported in the fifteenth century. Royal ordinances in time were passed favoring the improvement of the Merino, and great progress has been made in that direction. The number of Merinos in Spain is estimated from four millions upward. The native sheep of France were coarse, ill-formed animals, but in 1786 the Government purchased 376 sheep, selected from the best flocks of Spain, and placed them at Rambouillet, in the neighborhood of Paris, where there was an establishment devoted to breeding of animals. George III, in 1791, introduced the Merino into England; but although found to improve in size of carcass and in other particulars, they had given place to the true English breeds, because found less profitable. The "middle wools," embracing the Southdown, Norfolk, Dorset, Ryland, Cheviot, and others, are famous for their mutton. The Cheviots are the most hardy sheep of Great Britain, among the improved breeds, and any one who would try them in New England would be a public benefactor. They thrive on bleak hill-sides and poor pastures, and their meat is excellent. The Southdown is a native of the chalky hills of Southern England, on which grows a short, nutritious grass, well suited to mutton-making. By skilful breeding they have been brought well-nigh to perfection as regards shape, and their meat is most prized, combining as it does fatness with tender, lean meat, and having a flavor equal to the Highland mutton.

One hundred years ago, Mr. Bakewell, of Dishley, England,

undertook the improvement of the Leicesters, and created a magnificent family known as the Bakewell, or Dishley sheep. It was his aim by careful selection and breeding to combine, if possible, fineness of bone, beauty, symmetry of form, and tendency to fatten, with weight of carcass and a good yield of wool. His success is shown in the fact, that while he let his first ram for 17s. 6d. in 1760, he got in 1789, for one single ram, 1,000 guineas, and cleared $30,000 in that year by letting his rams.

Beside the sheep, Mr. Gold had samples of wool of all breeds, which he exhibited to us, and a number of engravings of famous sheep, taken from various works.

NINETEENTH DAY.—Feb. 22, 1860.

We have had to-day a very interesting session, the several lectures being replete with good points, and some of them especially worthy of consideration. The lecturers were, severally, Mr. Flint, on the Dairy Business; Mr. Gold, on Sheep, and Professor Silliman, Jr.

Milk, said Mr. Flint, as the first product of the cow, is composed of an oily substance, which gives it its richness; of a caseous, or cheesy substance, which gives it its strength; and of a serous, or watery substance, which makes it refreshing as a beverage; with a small percentage of sugar of milk, to which it owes its sweetness, and a slight proportion of alkaline substances, to which are due its medicinal properties. Under the microscope, it appears to be filled with myriads of little round globules, which float in the watery substance, and which rise to the surface in the form of cream, the largest particles rising first, and being the richest in butter. These globules are the butter particles, surrounded with a cheesy film, and the object of churning is to break this film, or coating, and to disengage the butter particles. The different constituents of milk separate on account of a difference in specific gravity. Milk will ordinarily produce from ten to fifteen per cent. of cream, though it is sometimes much richer than this, and twenty-five per cent. is sometimes, though rarely, obtained. The product in cream is more regular in several different lots of milk than the butter product which can be obtained from that cream. Caseine most resembles animal matter in composition and in nutritive qualities. The richest and most delicate butter is made from cream which has not stood long on the milk,—the cream that rises first making a far sweeter and better quality of butter than that which has stood a long time. If the milk is set in a favorable position, on shelves some feet from the bottom of the milk-room, around which a circulation of pure air can be had, from twelve to eighteen hours, in summer, is sufficient to raise all the best of the cream; and all that rises, under ordinary circumstances, after twenty-four hours, will deteriorate the quality to a greater extent than it increases the quantity. This is an important practical point, and ought to lead to the most careful experiments on the part of dairymen, who have been accustomed to let their milk stand for thirty-six and even forty-eight hours. An ordinary house-cellar is very rarely a suitable place to set milk, and it should never be set on the bottom of a cellar, if it is to raise cream. The bad gases (carbonic acid, and others, perhaps,) in the room, are near the bottom, and are apt to make the cream acrid. It will produce an inferior butter. The square box-churn is one of the best and most economical forms. To prepare new butter-boxes as quickly as possible, so as to make them fit to use to send butter in to market, or to the exhibition, dissolve common, or bicarbonate of, soda in boiling water, as much as the water will dissolve, taking water enough to fill the boxes, and at the rate of about a pound of soda for a thirty-two pound butter-box. Pour the water in upon it, and let it stand over night, and the box may be used the next day without fear of its tainting the butter. A delicate butter may be made by burying the cream in a cloth a foot deep in the ground, and leaving it for twelve hours or more.

Cheese has been used from a remote antiquity. Its varieties are almost infinite. This most important branch of American industry, the management of the dairy, involves the investment of a vast amount of capital, the aggregate profits of which depend largely upon individual judgment and skill; and any addition, however small, to the value per pound of the butter and cheese, would add vastly to the material wealth of the dairyman, and of the country at large. These articles are generally the last of either the luxuries or the necessaries of life, in which city customers are disposed to economize. They must and will have a good article, and are ready to pay for it in proportion to its goodness.

The great nicety and patience required to produce a first-rate quality of butter and cheese, and the gradually-increasing aversion of our farmers' wives and daughters to manual labor, have caused, in some districts, the butter and cheese dairies to give place to mere milk production; and sometimes low prices and cost of transportation to market have prevented the farmer from realizing a profit. Poor butter is at all times a drug in the market, and as the best can only be got by the most careful painstaking, Mr. Flint suggested that by imitating the "Dairy Associations," or "*fruitières*" of the Swiss Cantons, New England farmers might largely increase their profits at small risk. In the Western Reserve, there already exist cheese manufactories, or establishments, conducted by private individuals, for which all the milk of a large district is curdled and supplied at a stipulated price. The plan is said to have proved successful, and is found to be a public convenience. That part of the Swiss plan which Mr. Flint thinks best worthy of adoption in New England, is, to establish at a central point, in a village or neighborhood, a dairy establishment, under the charge of a thoroughly skilful overseer and trained assistants, supplied with all manner of improved presses, vats, churns, and other paraphernalia,—the completeness of the outfit being regulated by the amount of business to be done. This might be made by a joint stock association, or private individuals; the former being preferable, for a single proprietor would aim to get his curd at the lowest possible price, whereas under the joint stock plan the cost of manufacture is lessened and the common profit increased. The dairy furnishes to all subscribers rennet of the best quality, and requires them to follow a certain dairy management on the farm. At regular intervals the wagons go about to collect the curds, and the farmer gets his pay either for them, or for the cheese sold. In like manner, the cream could be sent for conversion into butter. Or if skim-milk cheese and butter were both made, both cream and curds would be sent to the central dairy. Allowing the practicability of this plan, and I can see no great reasons to the contrary, its manifest superiority is, I think, apparent. The dairy would become so famous for superior butter and cheese, that an extra price could always be obtained for them in market. In the Canton de Vaud, the butter made in these dairy establishments actually commands in market from one-fifth to one-sixth more per pound than that made at the small farms about; and in our country, where private wealth is more evenly distributed, the difference would undoubtedly be greater. Mr. Thomas Mottley, Jr., the West Roxbury breeder, gets fifty cents per pound for his Alderney butter in Boston, a fact which sufficiently shows that there are plenty of persons ready and willing to pay an enormous price for a superior article.

The care of sheep formed the subject of the lecture of Mr. GOLD. It should always be the object of the flock-master to keep his sheep in a thriving condition. The quality of the wool, as well as its quantity, and the general productiveness of the flock, demand this system.

Shelter is the first necessity in providing for wintering sheep successfully. The Southdowns will bear exposure better than any other class of sheep. The open fleece of the long-woolled parts on the back when wet, and admits the water, which completely drenches the animal, so that his abundant fleece is no longer a protection from cold.

Economy in feeding demands shelter for all sheep, as not only less food is required, but also, it is better preserved from waste. Water-soaked hay, or that which is in any way soiled, is always rejected. The improvement in the quality of the *manure* forms another argument in favor of shelter. That this is not only healthful, but grateful to the sheep at all seasons of the year, we see in the fact that even in summer they will seek their winter sheds at the approach of a storm, if they are within their reach.

Ventilation is of paramount importance, as connected with shelter; and to insure this, sheds, open to the south, are to be preferred. A stable with an open window will answer for a very small number, but the crowding of a large flock in such a place affects the organs of respiration, and may result in serious disease, and should never be tolerated.

The best form of *rack* has posts three feet high in the corners, a bottom of boards, the sides and ends of two boards each, the lower one the widest, with narrow perpendicular strips nailed on, to keep the stronger sheep from crowding the weaker. The spaces are larger in their perpendicular than their horizontal opening. The size of these, as well as the width of the rack, must be in proportion to the size of the sheep. Not more than one hundred of the fine-woolled sheep should be confined in the same yard, while the long-woolled will not thrive with more than twenty-five. A *hospital*, snug and comfortable, should receive any sheep that may be weak from age or disease, till, by careful feeding and nursing, they can be returned to the flock.

It is the worst possible practice to allow the sheep to fall away in flesh as the grass fails in autumn. The increasing wool conceals the shrinking carcass, much to the disappointment of the careless flockmasters. Better confine them in the yard than allow them to ramble about in search of some field of winter grain, which furnishes a little green food, but too light to be of any real value.

Winter fodder should embrace, in addition to the dry food, a due proportion of that which is green and succulent. Fine early cut clover hay, well cured, or that from old meadows, consisting of a variety of grasses, forms the best dry fodder. *Economy* demands that its quality should be good, else much waste ensues; yet the sheep is very fond of variety, and almost all of the so-called weeds become choice morsels. The botanist knows full well that a sheep-range will be most barren of the objects of his search. The immortal Linnæus tested the plants indigenous to Sweden by offering them, fresh gathered, to the various domesticated animals.

Horses ate 262 species, and rejected 212; cattle ate 276 species, and refused 218, while sheep took readily 387, and refused only 141 species. For fattening, add to the hay, roots, and grain, linseed or cotten-seed meal. The English system of winter feeding on turnips in the field is here prevented by excessive cold. Use them in the yards in moderate weather. Sudden changes from green to dry food, and the reverse should be avoided. Regularity in the hours of feeding is very important.

The amount of fodder varies with the kind of sheep, though it is not directly proportioned to the live weight. Ten small fine-woolled sheep will eat as much as a cow, the larger ones requiring more. 2 to 2½ or even 3⅛ per cent. of the live weight in hay value, is estimated by different authors as daily required.

No other animals except calves should lie in the yards with sheep. The losses from the horns of steers and the heels of colts more than balance any supposed gain. As the breathing of the sheep on the hay does not of itself render it distasteful to cattle, it may be gathered from the racks and fed in another enclosure.

It is estimated that 300 pounds of good hay will winter a small sheep, while larger ones may take three times the amount.

Water is absolutely necessary to the thrift of sheep in the winter. It is best brought into the yards, as the steep banks of streams prove dangerous to the sheep.

Salt may be provided in winter by a moderate salting of the hay two to four quarts a ton; but excessive salting must be avoided, as on such neither sheep nor cattle will thrive.

As the lambing season approaches, snug quarters must be provided for the breeding ewes, where they can be clean, warm, and dry. They will seek the necessary seclusion in the open field.

The increase from a flock of Merino or Saxony ewes, which rarely twin, may be from 80 to 100 per cent., while in the Southdown or Cotswold, 150 per cent., or even more may be raised.

Little can be hoped from legislative action as a protection from dogs. Bells attached to the necks of a few sheep in each flock deter the cowardly curs, or give warning of their attacks.

Sheep washing, shearing, and rolling the wool demand careful attention. Diseases come mostly from carelessness, and prevention must be our resource. The age of the sheep is determined by the teeth, but such irregularities arise in these as well as in other animals, that the Connecticut State Agricultural Society have decided to receive satisfactory testimony as to the age of any animal, rather than to depend on the indications of the teeth.

Of the three breeds on the stage, for the food consumed, the Merinos yield the most wool, the Cotswolds the most mutton, and the Southdowns mutton of the best quality.

The celebrated experiment of Lawes and Gilbert in England on 50 sheep, of each of the most celebrated British breeds, proves the Cotswold as giving for the food the most wool and mutton; the Southdown the least; yet, sold in Smithfield, the Southdown brought three cents per pound the most, so that the results as to profit were equal.

The Southdown is eminently fitted for the light lands of New

England; and when sheep husbandry shall have attained its proper place, it will be found as a chief instrument in that result, and their flocks will cover a thousand hills.

Prof. B. SILLIMAN, JR.'s second lecture on Meteorology was devoted to a description of the barometer, in its various forms, and the practical rules derived from its observation, applicable to the business of agriculture.

He first illustrated experimentally the discovery of the barometer, by Torricelli, in 1643. By means of an air-pump and two barometers, one in and the other out of the vacuum, he illustrated the influence of the atmospheric presence at the height of the mercurial column.

The model of the mercury barometer, made by Green, of New York, after the directions of Prof. Guyot, of the Smithsonian Institution, was exhibited, as well as other forms of this instrument.

He alluded to the practical objections to the mercurial barometer as an instrument for general use—its cost, if well made, and its unavoidable delicacy and fragility,—which must always act as a bar to its general use by the farmer.

Fortunately we had, in the "aneroid" barometer, an instrument free from these objections. Sufficiently cheap, not liable to be disordered easily, and withal sensitive and accurate enough for the use which is made of the barometer as a "*weather prophet.*" He proceeded to give a popular description of the essential features of the aneroid barometer (or barometer "without a fluid," as the term implies). This instrument was invented by Mr. Vidi, of Paris; it is without mercury, and consists of a flat and circular metallic box, the cover of which is very thin and corrugated, or in ridges and furrows, concentric with the walls. The air is exhausted from this box, which is then hermetically sealed. The result is, that the elastic cover rises and falls with every change in atmospheric pressure. By means of a combination of levers and springs, these movements are communicated from the centre of the cover to a pointer which moves over the graduated face of a card, on which inches and hundredths are inscribed. The whole apparatus is encased in a brass box, about four inches in diameter and two inches deep, covered with a front glass, and resembling in general appearance a chronometer case.

These instruments are now made by Mr. E. Kendall, of New Lebanon Spa, N. Y., well known everywhere for his mercurial thermometers. His instruments compare well with the French, and with the movements of the mercurial barometer, and sell for the moderate price of ten dollars, or one-third the cost of a Smithsonian barometer. Although for purposes of scientific accuracy nothing can replace the old form of mercurial barometer, Prof. Silliman did not hesitate to recommend the aneroid as the best barometer for the use of the farmer. Numerous testimonials, from farmers who had used them, showed their utility in enabling the farmer to choose the time of cutting and curing his hay, planting, &c.

Prof. Silliman explained why the words "fair," "changeable," "foul," "tempest," &c., &c., written on the scale of the cheap forms of mercury barometers were entirely unreliable. It was only at the sea level that the barometer stood at an average height of thirty inches, and hence a mere change of place, rising a few hundred feet, would make the barometer fall permanently below "*fair weather*," whatever the face of the sky might say to the contrary. That the use of the barometer might be better understood, he enumerated the following general rules, which embody the results of long and various experience in different places :

1. When the mercury is very low, high winds and storms are likely to prevail.

2. Generally the rising of the mercury indicates the approach of fair weather; the falling of it shows the approach of foul weather.

3. In sultry weather the falling of the mercury indicates

coming thunder. In winter, the rise of mercury indicates frost. In frosty weather, its fall indicates thaw, and its rise indicates snow.

4. Whatever change of weather suddenly follows a change in the barometer, may be expected to last but a short time.

5. When the barometer alters slowly, a long succession of foul weather will succeed if the column falls, or of fair weather if the column rises.

6. A fluctuating and unsettled state in the mercurial column indicates changeable weather.

In these rules, the "index of the aneroid" may take the place of "the mercury column."

Prof. Silliman called to witness the experience of Mr. Jos. Lesley, Jr., of Phila., one of the class who had, as a topographical engineer, made great use of the aneroid as a levelling instrument. This gentleman stated that he had used this instrument during the whole season in determining contour lines over hundreds of miles of broken country, and had found, on calculating his lines at the end of the season, the differences quite inconsiderable. He was disposed to rank the aneroid, as an instrument for scientific uses, higher than Prof. Silliman had placed it, but stated it was important to apply always a correction for temperature—a sort of "personal equation," varying for each instrument.

Prof. Silliman concluded by quoting still farther some of the general conclusions of Prof. Henry, Prof. Coffin, Mr. Espy, and others, as embodied in the Agricultural Reports of the Patent Office and of the Smithsonian Institution. He strongly advised the class to study the articles on meteorology, contained in the documents for the years 1856 to 1860, as being far the most reliable of anything hitherto within the reach of the general reader.

In the evening there was delivered a lecture by CASSIUS M. CLAY, on stock and stock-breeding.

Mr. Clay's first lecture was given in the Baptist church,

to an audience of several hundred persons. He commenced by stating that he had come here as a progressive farmer, to lend his aid and influence to a movement which he deemed of great importance, and the necessity for which he had for years appreciated. We hear it said on every hand, and especially by politicians, that farming is a respectable business; but he thought that no amount of honeyed phrases or plausible talk would make any calling respectable. Agriculturists were ahead of most others in moral and physical developments. If farmers would be really respected, they must refine and cultivate themselves into respectability, and not wait for it to be done by others. They must carry their capital into the country, and use it judiciously in advancing their farm practice. Taste should be cultivated; and rural architecture, landscape gardening, and other things which render a country attractive, should especially be fostered. To further this great object this Convention had been called, thanks to the sagacity and enterprise of Prof. Porter; and although it would have been perhaps more convenient to him (Mr. Clay) if it had held its session in Kentucky, yet, it being in Connecticut, he was willing to come hither, for what tended to promote the advancement of New England farming was as dear to his heart as if it were especially pointed at Kentucky interests. It is the sheerest madness for farmers to drain the heart of their farms and invest their funds in stocks and bonds, for the application of capital to farm improvements would give as large comparative profit as it would in any other business. The introduction of better classes of farm stock, Mr. Allen had told us, would add from forty to sixty millions of dollars annually to our wealth. If we took this sum for a few years and applied it to farm improvement, what magnificent results would be attained! Through the interior of Kentucky the farmers were so sensible of the profit derivable from improved stock, that they would no longer purchase common scrubs at any price, nor even give them standing room on their farms. For they had found, and others would find, that to purchase them at any price was in the long run poor economy.

He would not attempt to describe all the multifarious breeds of cattle in the civilized world, but would confine his remarks to the five leading British breeds—the Alderney or Jersey, the Ayrshire, the Devon, the Hereford, and the Shorthorn. In size and weight the Alderney is the smallest; it is supposed to have come from Normandy, but has been improved in the Channel Islands, and is greatly superior to what it formerly was. It is a picturesque-looking animal in appearance, rather than a strictly beautiful one. Those which he had seen were mostly ewe-necked, sway-backed, high in the withers, full bellied, and narrow in the girth. But he understood that by skilful breeding there had been many individuals of the breed made up to a symmetry and development quite creditable. The Alderney, he conceded, gives the richest of all milks, but little in quantity. Taken to the country, it was an active animal, capable of getting a living on scanty pastures. It will thrive in some degree almost anywhere with us, but undoubtedly does best in districts which are the same isothermally as its native land.

The Devons are supposed to have been brought to England with the Celts, and are, perhaps, rightly regarded as the oldest breed of the British Isles. They are mostly a dark red, with close, fine curly hair. They are a degree larger than the Alderney, are heavy in the head and horn, do not carry out the rump well, but are a very good animal withal. They give rather more milk than the Alderney, and of almost as rich a quality. They are not very heavy in the brisket, and, being narrow between the shoulders, are enabled to move briskly, and are thus adapted to working under the yoke, although rather light for heavy draft—and hence they have been improved by a cross of the Shorthorn for oxen. The Longhorns have been tried in Kentucky, but abandoned, for they did not prove either famous milkers or feeders. The Devon is too small for Kentucky, and for other districts like it where there is abundance of heavy, rich pasturage. They do not aim at getting single famous milkers in his State, for they keep many animals, and a little milk from several is fully adequate to their purpose, beef being the great end.

The Hereford he does not deem an original breed, for they were formerly of a dun and dark color, and are now white-faced and throated; a peculiarity which he thought owing to a cross with the Glamorgans, and not the Somersets. Their greatest inferiority was that they were miserable milkers; a very bad fault, for there are doubtless a thousand persons who wish a milking animal to one who wants to make beef. The Hereford, as compared with the Shorthorn, is coarser in the shoulder and thicker in the hide, beside wanting that general symmetry which characterizes their great rivals. A good handling quality of hide is highly prized by the butcher, for a mellow, spongy skin indicates a good quality of beef, and that well "marbled." In this important feature he had found the Hereford deficient. He was aware that this breed is a favorite with butchers, but thought it greatly due to the fact that it lays on its fat in patches on the inside of the carcass, and thus goes in the "fifth quarter" as the butcher's perquisite.

The Shorthorn he deems an original, and not, as popularly supposed, a created breed. They vary much, it is true, in color, but these variations are well defined, and evermore repeated. He had never seen a real Shorthorn without some patch of white on it. The physiognomy of the race is the same as in olden times; a fact which he thought demonstrated in their resemblance at this day to the outline of an old Shorthorn cow sculptured centuries ago upon a marble slab in an old church at Durham. The Shorthorn has not only perfection of form, but size, fattening properties, and milking qualities as well. In England, Scotland, and this country, any dairy which is famous will generally be composed of Shorthorns, either thoroughbred or grades. We may breed out the milking quality, but we may on the other hand develop it to a great extent by careful breeding from milking families. He has had an animal give thirty-two quarts of milk daily;—the Shakers of Kentucky report one giving forty quarts a day—and he believed the breed will make more butter and cheese than any other. In early maturity they are unrivalled. At two years old they have been sent from Kentucky to the New York market in prime condition, though three and upwards is the usual age.

He was not of those who admitted that the improved Shorthorn family had been created by Charles and Robert Colling, for Colling himself admitted that he had bought fine animals wherever he could find them before he began to breed for himself, and Phœnix and Lady Maynard were as fine animals as he ever bred. He had bred judiciously, and improved the breed in extent, but its origin must be sought prior to the days of Charles Colling's Hubback. Perhaps it may not be advisable to use them in New England to the exclusion of other cattle; but throughout the whole interior of this country, where the climate is fair and the pasturage good, they would, as they had in Kentucky already, run out any other of the leading breeds which might be placed in competition with them.

The Ayrshire is essentially a modern breed. At least there was no such breed famous in Ayr a hundred years ago; and he was of the impression that it had originated in a cross of the Shorthorn with the West Highlanders. It has many of the characteristics of the Shorthorn; is, next to it, the heaviest feeder; and its great milking properties he thinks due to that part of its parentage. Carried to poorer pastures in England and elsewhere, the Ayrshire does not thrive as well as on its native fields. Some public-spirited farmers in Kentucky have recently imported some of the breed, and will give it another fair trial; but Mr. Clay believes the same unfavorable result will follow as has heretofore.

Mr. Clay claimed that his favorite breed possessed all the essential points of true beauty. Beauty, he thought, was composed of five elements. 1st, Propriety: that is, the adaptation of means to an end. The full impression of beauty is never conveyed to the cultivated mind if the eye is shocked at seeing an unsuitableness of form to the purpose in view. 2d, The elliptical line, or the oval. We make our picture-frames oval because that is the most beautiful shape, and so do we our plats of grass and the leading features of a landscape garden, while the female face is never absolutely faultless unless it presents the oval form when viewed in front. The Greeks made the face oval in the Venus, but fuller in the forehead in the Minerva and Jupiter. 3d, Color. The brightest gems are the best, and the greatest luxuriance of tints is lavished by nature, where she makes her loveliest handiwork. 4th, Smoothness of surface. The angular form is not admissible in a connection with the beautiful; and roughness is merely angularity infinitely multiplied. 5th, Proportion, or the harmonious arrangement of parts. All these qualities he thought combined in the perfected Shorthorn of our time; and we are bound to respect the beautiful, for we spend at least ten times as much for it as we do for the purely utilitarian.

Mr. Clay illustrated his remarks with the aid of a large painted sketch of one of his Shorthorn cows, which was suspended at the back of the platform. He was loudly applauded on resuming his seat, as also was the announcement by Prof. Porter that the second lecture would be given to-morrow morning.

Mr. Clay being limited to one hour and a quarter, by agreement with other lecturers, did not go as fully into the description of the several breeds as he had desired.

TWENTIETH DAY.—Feb. 23, 1860.

To-day two lectures on stock-breeding were given, one by Mr. Allen, the other by Cassius M. Clay. There was no great diversity of opinion between the two breeders as to the broad, fundamental laws of the art; so that while I am debarred from giving sketches of both lectures, their substance can be as well condensed into one.

Mr. Allen read a letter from Mr. John T. Norton, the famous Jersey breeder, of Farmington, Ct., which embodies so much valuable information, that I cannot refrain from publishing it in this connection. Mr. Norton says:

"The pure Alderney cattle come mostly from the Island of Jersey, in the British Channel, where they have been kept free from mixture for a hundred years,—no other breeds being allowed on the island. Similar cattle are found on the other Channel Islands; but all more or less mixed with other breeds. About two thousand head of cows and heifers are annually sold from the island, the area of which is not much greater than that of one of our largest New England towns, at an average of £5 sterling each, making £100,000 sterling, or $500,000, from this source alone.

"The Alderney cows are small and thin, with delicate deer-like limbs—generally light yellow or fawn color—always poor in flesh when in milk, but taking fat readily when dry. They are remarkable for gentleness and docility—easily kept, and usually give milk nearly up to the time of calving.

"The important question in relation to these cows is *their value compared with other breeds.* It will be conceded at once that for *fattening*, for *labor*, and for *furnishing milk for sale*, they are inferior to almost all other breeds.

"In Great Britain they are kept mostly by the wealthy, to supply their own tables with milk, cream, and butter. Colman says: 'Every nobleman and large land-owner keeps one or more tethered on his lawn, for family use.' They are also kept by many London dairymen in the proportion of one Alderney to ten other cows, to color the milk for market.

"My own experience, after many years, has led me to the conclusion that for *butter-making* they are superior to any others, yielding more in quantity and of better quality.

"In all other breeds, and also among grades, superior milkers and butter-makers may be found, equalling in quality of butter, and giving more milk, and producing more butter, than most Alderneys. But there is no other breed known here that can always be relied on. I have never known an Alderney cow whose milk and butter had not the characteristics of the breed. They differ, as do others, in quantity, and somewhat in quality; but the peculiar color and quality are manifest in all.

"The daily yield of milk of each cow, during their best milking period, varies from six to twelve quarts. This milk will make about one pound of butter to six quarts of milk. One pound from twelve quarts is not far from the average yield from other breeds.

"The average product of butter from my cows in 1859, was a fraction over two hundred pounds each. The average product of the dairies of the State of New York, I think, is about one hundred and twenty pounds to each cow.

"The premiums by the New York State Society for the greatest product, have been given to dairies producing about one hundred and eighty pounds each cow.

"My cows have had no extra feed. In summer they are kept on grass only. In winter they have one feed daily of cut corn-stalks, straw, or coarse hay, with a slight sprinkling of bran, or cotton-seed meal, and two feeds of dry hay.

"The average price for which my butter sold in 1859, was thirty-five cents. The price now is forty cents. In March and April, it is to be forty-three cents, by contract, in Boston.

"In relation to any improvement in the stock, I am of the opinion that none can be made by crossing with any known breed. Increase in size, or an increased disposition to fatten, will be gained only at the expense of a loss in cream and butter.

"An analysis of numerous specimens of milk made in 1858 by Dr. S. R. Percy, under the direction of the New York Academy of Medicine, resulted as follows, viz.: The milk from six of my Alderneys, taken indiscriminately, exhibited butter compared with the *best* other milk, as seventy-two to forty-seven, and compared with *mixed* country milk, as seventy-two to forty.

"I am yours, very respectfully,

"JOHN T. NORTON."

Mr. Clay commenced his second lecture on Cattle Breeding by pointing out, on the large sketch of a cow, the several good and bad points of the improved Shorthorn. There should be no surplus meat about the head, for it is all waste, or nearly so, and it consumes a quantity of food in being created which might be more profitably employed. A large dewlap, being poor for meat, and the skin inferior for leather, and a useless deformity, should be avoided. A straight spine indicates a state of health, as well as fine beef. Whenever an animal is too closely bred, or suffers in health, the spine droops, and the animal is called "sway-backed." The girth should be as large as possible, for just under and behind the shoulders are located the vital parts—the heart, lungs, &c., and ample space should be given to them for full development. Without this there can never be the perfection of vigorous growth and hardiness of constitution. The ribs should be joined to the spine at, or near, a right angle, should spring well outward, and drop well down toward the belly,—that there may be capaciousness of carcass to hold the viscera and food. The rump should be long to hold fine meat, and a long stretch from hip-bone to hock is necessary to give powerful leverage to working-oxen. A large brisket, projecting forward, and dropping below the line of the belly, he does not like, but rather aims at getting one of medium size, which indicates a strong constitution. A too large one is a deformity, a too small one a sign of weakness; it is but a wall for the chest. When too large, it forces the animal to turn slowly, like a long ship, and makes rapid motion difficult. Breadth of chest is to be sought after, for manifest reasons. The flank should drop well down, not so much for the profit it gives as to preserve a general symmetry of form. The beast should be well ribbed back. That is to say, there should be little space between the last of the short ribs and the hipbone. If an animal is too long in body, it is apt to sway, or sink in the back, on the same principle as a long rope stretched from two points sinks at the centre. The feet and legs should be small, though not weak. The shin-bones make fine soup. In Kentucky, they esteem as peculiarly delicious a part which we throw away, viz., the feet. They first parboil them until well cooked, when the hoofs come off. They are cooled, and then reboiled, and before being served up, cream is added, with chopped onions, and some pepper and salt. Mr. Clay said he would travel further to get a dish of feet than a bowl of green turtle soup. I think we had better get our wives to try it.

The loin should be broad and full—here is the prime beef. The tail set on a level with the back, and large—falling from well back, and tapering to the joint. The perfection of girth, therefore, in an animal is the perfect circle, filling up the crops well.

Twenty-eight years ago Mr. Clay began breeding Shorthorns, and imported the first thoroughbred into Madison county, Kentucky. He was a candidate for the Legislature at the time, and thinks he lost many hundred votes because he dared to pay $100 for a blooded bull. His neighbors thought it better to send him to a lunatic asylum than to the Legislature. Things are changed now. These very men come to him and pay sometimes $300 for a single animal. In former and more prosperous times he has had 500 or more animals feeding on his farm at once, and has handled as many as a thousand head in a year.

His herd is now small, but choice; for he has sold the poorer animals and kept none but the best. He breeds from the stock of 1817, and later brought, and holds his own with the owners of recently imported animals.

Breeding as an Art.—In breeding, we cannot be too strongly impressed with the fact that LIKE PRODUCES LIKE. Does a man gather grapes from thorns, or figs from thistles? We should regard purity of blood, choosing our breeding-animals from a family in which there has been a succession of animals of the same type. If we use a grade bull we are never sure but that the calf will take on the type of some one of the worst of his ancestors. Climate, soil, and food, have a great effect on the physical development of both men and animals. A genial climate and abundance of food make beautiful and healthy animals, and the magnificent Shorthorn doubtless owes it supremacy to the fact that it had both of these aids in the valley of the Tees.

We should strive to breed so that the defects of one parent may be counterbalanced by the points of the other. If the dam is inferior in girth, the sire should be fine there; if the one be too long in body, the other should be rather short. We should never cross animals of very great dissimilarity of development, however, lest the defect be thereby unreached, nor should such diverse breeds as the Alderney and Shorthorn be mingled. Mr. Clay is a decided opponent to the practice of "in-and-in" breeding, basing his objections on what he deems adequate experience and observation. In his opinion it is as wrong to breed closely with animals, as for cousins and other near relatives to intermarry. Bakewell, of Dishley, England, proved that fully. He gathered the best specimens of sheep and Longhorns, and bred them up to good specimens—making the Leicester into the improved Dishleys, and very superior Longhorns. But by "in-and-in," or close breeding, the stock ran down. The Bakewells, or Dishleys, had to invigorate with new crosses, and the Longhorns, being at best a poor breed, have gone to nothing! He referred his audience to a full discussion of the subject of "in-and-in" breeding, between himself and others, in the *American Agriculturist* for 1859.

As a general rule the female should be comparatively larger than the male. Mr. C. had found it very hard for scrub cows to be delivered of the fœtus by a large Shorthorn bull. A large coarse bull is especially to be avoided.

The whole art of feeding might be summed up in the remark, that the animal should never recede in flesh till mature, but be kept in good growing order always; never too fat nor too lean. That is the way to have perfection of form—other things being equal. When animals are grown it is not so important to keep them always in good flesh; although he has known show animals, once too fat, ruined in health by getting too poor! Too much fat will destroy the breeding power in male and female frequently. In Kentucky they are fast rivalling, if not excelling, England. Because, by the system of open stables and out-door exercise, the laws of health are better observed. The animals in England kept too much in stables and fed on heating food like oil-cake, have to be rowelled, bled, and purged! Of course we who follow nature's law, need none of that; and will ultimately beat them in perfection of form, &c. With us, in Kentucky, there is none of that degeneration of animals imported, which is so often talked of in the North; because we keep up the favorable surroundings and means of progress.

The Shorthorns will conceive at under four months. But Mr. Clay prefers to have them 2 years old before they are impregnated. If they calve younger they should be fed highly,—for, if they are not, the fœtus takes up so much of the nutriment, that the mother is stinted in food for necessary assimilation, and becomes stunted and ill-formed. Possibly early breeding may rather more favor the milking quality; but his experience is not sufficient to accede without further proofs to this general idea.

All breeds for permanent breeders should be thoroughbred. The Shorthorn brings up the native cattle wonderfully,—but they should be bred all the time to a thoroughbred bull, and the grades should not be bred, if it can be avoided, to any other bull. This way will bring a herd up wonderfully by the simple outlay for a good bull.

With regard to color: within the bounds which mark a breed, he knows no utilitarian color. The Shorthorns combine red and white in all proportions; but no other color, except yellow, is admissible. Red is just now the favorite color. Roan was once, and may be again. White winters and fattens as well in Kentucky as any other color. Some of the finest bullocks ever sent to the New York market were grazed by him, and were whites. The finest and best fatted heifer he ever saw was descended from the 1817 stock of Shorthorns, and was white—weighing over two thousand pounds!

Mr. Clay did not believe the doctrine that the features of the first sire were impressed to some extent upon all succeeding fœtuses. He thought that idea had been originated by the women! Mr. Clay thought we were in the infancy of the art of breeding—full of uncertainty now; yet the laws of breeding were as fixed as the laws of Physics. All we wanted was knowledge. We knew no way at present of influencing the sex—though he thought the most vigorous animal influenced the sex. He thought, if an old bull went to many cows, the calves would be heifers mostly;—but if a young bull went to a few and rather old cows, the result would be males. We needed more intelligence and more close observation. The course of higher progress was in such efforts as those now here making. Let our motto be—Excelsior!

TWENTY-FIRST DAY.—Feb. 24, 1860.

According to the pre-arranged schedule, we should have had a lecture from Mr. Donald G. Mitchell (Ik Marvel), on Rural Economy, and two from Ambrose Stevens, on Horses; but Mr. Mitchell excused himself on the ground that his subject had, in great degree, been anticipated in preceding lectures, and owing to some fault in the mails, or otherwise, Prof. Porter's letters and telegraphic dispatches failed to reach Mr. Stevens. We have been in both cases disappointed; for there is no such graceful pen as Ik Marvel's enlisted in the cause of agriculture, and Mr. Stevens is regarded as one of the best-informed and scholarly of our horse and cattle breeders.

Mr. Mason C. Weld, a pupil of Liebig's, and now one of the editors of *The Homestead*, gave us last evening a sensible lecture on Agricultural Associations.

After remarking upon the general benefits of association among farmers—the proposition being maintained that in proportion to the degree of enlightenment attained, is the readiness of individuals to communicate their knowledge and experience for the benefit of others—Mr. Weld took up, separately, the various kinds of organizations sustained for mutual benefit among farmers. Cattle insurance companies, on the mutual plan, were passed with simply calling attention to them as having a very beneficial effect in necessitating accurate veterinary knowledge and practice, and the humane treatment of poor, ailing beasts, instead of the barbarities now too often practised. Agricultural associations were treated under the following titles: Temporary Farmers' Clubs, Permanent Farmers' Clubs, Town Clubs; County, State, and National Agricultural Societies.

The Temporary Farmers' Clubs are simply meetings of farmers—e. g., those attending a fair, or members of a State Legislature—who assemble, appoint a chairman, and talk agriculture. The requisites to success are—1st, short speeches; 2d, an active, prompt chairman.

The Farmers' Club proper, is an organization—the simpler the better—of the farmers of a neighborhood. It was advised to have, in general, no regular constitution, but a few simple rules instead; to elect a presiding officer at every meeting, but to have a permanent secretary, with extraordinary powers, appointed annually. The primary object of the farmers' clubs is, to promote, in every feasible way, the improvement of the agriculture of the district. This is accomplished by making common stock of the knowledge possessed by each member; collecting statistics; keeping a record of extraordinary events; distributing seeds and grafts; testing implements; aiding each other by counsel; maintaining regular meetings; a library, &c.

A plan for breaking up the boys' debating-society system, which such clubs are apt to fall into, to the disgust of good farmers, and the ultimate discontinuance of the clubs, was prefaced as follows: Suppose the clubs to represent fairly the best farmers of their districts, and to meet all of them (that is, all of the State or county) upon the same day, about the first of each month. A set of questions for each month in the year being set forth by the central State association, each farmer may answer each question as concerns his own farm; and as the questions should be carefully prepared with a view to develop the most important facts and statistics, a summary of the answers of all will give a view of the position of the town, prospectively and retrospectively, as regards its products seeking a market; sales and purchases; crop prospects and results of harvests; increase of stock; diseases among domestic animals; prevalence of disease among crops; insect ravages, &c. The plan is, that these monthly statistics should be placed on file; a summary sent to the secretary of the county, or State society, as soon as possible, in order that the more important facts, affecting the market, may be made public, while all should be kept on file at one place or the other, for reference and investigation. The object to be gained is the personal interest of members in the club, and especially of all good farmers, and the full accomplishment of the legitimate ends of the association. The Farmers' Club was held to be the most important means of educating a class of energetic and intelligent farmers, to whom may be intrusted the affairs of the State and County Agricultural Societies.

The County Society should be made up of the Farmers' Clubs, and the two classes of organizations should work harmoniously together, each doing its own work. A more definite organization is needed—officers elected for one year at least, a vice-president, or director, being chosen from each town by the Farmers' Club of the town. The fairs were shown to be a chief means of carrying forward the objects of these societies, and also the great desirableness of, and the great difficulty of securing the services of fair, honorable, intelligent, reasonable men to act as Awarding Committees. The cure for the state of things now commonly existing lies in first offering fewer premiums, and increasing their value; second, allowing no discretionary premiums, or gratuities, to be given in classes in which regular prizes are offered; third, insisting that the award shall represent the accurate estimation of the committee of the worthiness of the animal, or article, without regard to the encouragement or reward of the owner for making the exhibition; fourth, throwing the whole of the responsibility of making a correct judgment upon the committee, and securing the fairest and best men. Offering prizes for articles of no agricultural use or importance, as well as making balloon shows, ladies' riding-matches, &c., were condemned as undignified and unworthy of an Agricultural Association.

State Societies should—as most do—depend upon the county organizations, as these in turn do upon the clubs; and their management is much the same—only upon a larger scale. Museums of all things of an agricultural bearing, implements, grasses and grains, seeds, models, &c., and libraries of home and foreign journals and books of reference, were advocated, as well as the practical use of interchanges of seeds, grafts, &c., through the medium of this mutual dependence of the societies and clubs one upon another.

In conclusion, the lecturer advocated strongly the establishment at once of an experimental farm, in connection with a thoroughly furnished laboratory, referring to the debt the world owes Lawes and Gilbert for their experiments at Rothampstead, and to the most weighty results developed by the investigations in France and Germany, which latter country has now in operation more than forty experiment stations under the management of competent men of science in connection with practical farmers.

The convention assembled at 9 o'clock this morning, and listened to a lecture upon the methods in use for "Breaking and training horses," by Dr. DANIEL F. GULLIVER, of Norwich. The introductory part of his lecture was spent mainly in describing the characteristics of horses as distinct from other breeds of animals. Their high spirit, great intelligence, and susceptibility to fear as well as kindness, render them a proper companion for man. Upon these principles in his nature do the principles of training depend. The systems of Baucher and Rarey only will be discussed. The former is easy and extremely simple, even almost stupid when considered in its several parts, yet as a whole it is eminently successful. It was first introduced into this country in 1851, at Philadelphia. It is very popular in France, so much so that in the eight years from '41 to '49, nine editions were sold of that work. It should be owned by every man that owns a horse. This system is designed principally to finish the horse for the saddle; but its principles are applicable to *all classes of horses*, excepting those intended for heavy draught. Mr. Seth Craige, of Philadelphia, who has used this method even from the proof sheets, whence he learned it, says a horse fitted for the saddle makes the best harness horse, and this is the system for developing speed and harmony of action. The animal is well balanced, and so trots better because all its movements are regular. Everything is graceful, and all the forces of the system assist each other. Herbert says a horse has greater docility, as well as better style and action, if it be thoroughly trained in the saddle before being put into the harness. The improvement is proportioned to the extent and degree of supplying more or less. To break properly, we should seek out sources of resistance to graceful motion, whether from the physical nature, or from a previous imperfect motion. All the resistances we should overcome by a progressive system of suppling, applied successively to the principal muscles from the head to the haunches. The work is generally badly begun, and bad habits are produced. The term *force*, as we use it, is muscular power in action. The forces of the horse are subjected to control by giving to the body a new balance, where all the *instinctive* forces are changed *to transmitted.* Forces are termed instinctive, when the *horse* determines the use of them; transmitted, when the *man* determines the use of them. Any man who has a modicum of " horse" in his disposition, may go through the supplings with his beast and break him well, but it is a gradual process. The forces should be first conquered, and finally studied so as properly to direct them. The animal should be taught the exercise of the forces of balance and motion. The focus of these forces is the centre of gravity of the animal, while at rest or in motion. The various technical steps in this process were gone through with in detailed description by Dr. G., commencing from the head and passing along backward. The muscles controlling the action of the animal are to be subdued individually. Direct and indirect flexion of the jaw and of those muscles which join the head to the neck, is calculated to aid his intelligence, and thus secures, more satisfactorily, what the *bitting harness* is designed to accomplish. The very use of training is to supple the horse to the hand; hence, what need of machinery? The neck and head are the two props upon which the horse relies to resist efforts to break him. The whip should seldom be used as a punishment, but when it is, let it be done energetically and suddenly, to frighten the horse and intimidate him, rather than by being passionate and furious, to make him angry. There is great advantage in removing the support before shifting the weight. And an animal must be taught to lift his foot before throwing the weight in the direction he wishes to move. Whatever is possible with a horse, Baucher's system makes attainable, and this by the easiest means. The action of wild horses is perfect, but when domesticated, chained to a stable, holding sharp bits, (of which forty kinds were lately on exhibition in Philadelphia,) the native grace is lost; but all the high capabilities remain, and are seen in the common horse, when developed by the supplings of Baucher. "What is gracefully done is easily done," is a maxim as applicable here as elsewhere. The horse of the present day wears out too soon because its education is forgotten, and it is treated like a machine. Proper breaking and training would add 30, 60, and in special cases even 100 per cent. to the value of horses. Not one in a hundred, Herbert tells us, or even in a thousand in the United States, was ever properly broken, and not one in fifty has the proper rudiments of an education. In ten years the demand for saddle horses will be increased in a twelve-fold ratio. The foolish desire of the community is for speed. If a horse is not fast, he is good for nothing. The high prices these fast horses bring would be a fortune to some of the farmers. But they forget how many intermediate hands these prices pay. Most farmers try to breed something FAST (tempted by the fabulous price, or because their neighbors do); thus the whole community is involved, and the market glutted with a class of horses, which if they fail in speed are fit for nothing else. This process is the best to develop the animal, but it must be progressively and carefully applied. Such is Baucher's effectual means to annul and equalize all resistances. Rarey's method is applicable to all horses, of all ages, but to those especially who have never been handled at all, or have been badly handled. It consists in laying the horse flat upon the ground, by simple and easy means, though in extreme cases the terrible process of choking is resorted to. In 1858, the New York Tribune gave a compendium of it. The nigh fore leg is to be bent at the knee, and the hoof strapped to the leg. A long strap is fastened low on the off fore leg. Thus we have the horse on three legs and under control. After two or three throws he becomes entirely submissive, and no act of kindness is thereafter lost upon him. Affection to his master, personally, is the great result of the Rarey method. This throwing need not injure the animal, since it may be done with some soft material under foot, or the knees be protected by pads; moreover, the posture is one the horse assumes voluntarily whenever he wishes to lie down. In a herd of native horses, social position is determined by the varying degree of muscular force, and if so overcome by man, he will be convinced he is his superior and yield. If you go so far, you have now a pupil which will learn anything. But be patient, even-tempered, not hasty, and never angry. Rarey says anger and fear should not be known to trainers. Ask nothing that you do not want, and then always have it performed.

At the close of the lecture, Wm. Whittlesey, of New Britain, was called to the chair, and several questions bearing on different points, were asked the lecturer, all of which he answered satisfactorily.

THE COURSE CLOSED.

New Haven, Feb. 25, 1860.

After Mr. Mason C. Weld's lecture, as I yesterday stated, there followed an address by Professor Porter, and a sort of discussion upon the success of this plan of agricultural education. Professor Porter has modestly refrained from speech-making from the very commencement, and has stooped to none of those tricks to make popularity which the engineers of less important enterprises often employ. He deemed it incumbent upon him, at the close of his course, to give a sketch of its inception, and show what reasons he had to believe its permanence secured.

I shall make no report of his remarks, for in his preface to this volume he has stated his views at sufficient length, and much better than I could.

The Professor having concluded his remarks, Mr. H. A. Dyer, Treasurer of the Connecticut Agricultural Society, was, on motion, elected Chairman of the meeting, and Mr. H. A. Pitkin, of Louisville, Ky., Secretary; and an organization being thus effected,

Dr. Wm. A. Townsend, of Lockport, N. Y., offered the following preamble and resolutions:

Whereas, The Faculty of the Scientific School of Yale College have instituted a course of lectures, given by scientific and practical men, in relation to all the various departments of agriculture, combined with a system of discussion, questions and answers, statements and illustrations, we who have participated in these interesting exercises feel a desire to express to the agricultural community at large our views and opinions in regard to the same: therefore,

Resolved, That we cordially and fully approve of this method of diffusing and disseminating agricultural information, and regard it as the opening of a new era, and presenting new facilities to all classes of agriculturists in our country for obtaining correct and reliable information and knowledge, in relation to the cultivation of the soil, and therefore recommend this method to the candid consideration of all farmers and cultivators, of whatever age, position, and locality.

Resolved, That in view of the success of this Convention, the gratitude of the agricultural community is due to Prof. John A. Porter and his associates for the design so happily conceived, and his untiring efforts in carrying it out.

Resolved, That we entertain the hope and express our earnest desire that this may prove the germ of a permanent institution, endowed with all needful facilities for illustration, which, as a department of the Yale

Scientific School, shall greatly promote the cause of Agriculture, and elevate the farmer to his true social and intellectual position.

Resolved, That we hereby tender our warmest thanks to the citizens of New Haven, who have extended to us so many acts of kindness and hospitality during our sojourn here, and we beg them to be assured that the evidences of their generosity and goodness are duly appreciated and will long be remembered.

These were unanimously adopted.

Mr. M. L. Holbrook, of the *Ohio Farmer*, then offered an additional one, as follows:

Resolved, That the thanks of this Convention be tendered to the lecturers, for the promptness of their response to the call, and for the very able and faithful manner in which they have conveyed both scientific and practical instruction.

Mr. Lewis F. Allen said that he had had no doubts of the success of this plan from the very first, and he had come 500 miles or more to show his disposition to aid the movement as much as lay in his power. The names of the lecturers were of themselves a guaranty of the value of the course, if Yale College had not lent its influence toward it. In his opinion, Yale College, with all her great achievements, had never done anything so great or important as in establishing this Agricultural Lecture system.

Mr. S. B. Parsons, of Long Island, thought that an experimental farm would grow naturally from this movement, and if it did, and it were properly conducted, who could prophesy the national benefits which would result?

Mr. Judd, of the *Agriculturist*, gave his unqualified approval of the matter, and promised the aid of his paper to the fullest extent possible.

Prof. B. Silliman, Jr., said that when he saw all this enthusiasm and good feeling he could not help recalling the by-gone days of 1846, '47 and '48, when the late John Pitkin Norton and himself had, after much trouble, obtained the recognition of an Agricultural Department from the College officials.

How tremblingly they two had started on their work, with their little collection of apparatus! Their first class of pupils was very small in number, but all its members had achieved honorable reputations. Their beginning, small as it was, was still due to the enlightened views and generous enthusiasm of Mr. John T. Norton, of Farmington, who contributed $5,000 toward a fund to endow a Professorship of Agricultural Chemistry, and, since his son's untimely death, had allowed the income from that sum to remain for appropriations to the same end.

The CHAIRMAN spoke feelingly to the memory of Norton, and foresaw for this department a flattering future.

Other gentlemen expressed similar views, and, after a pleasant evening's discussion, the meeting adjourned, without day.

It is not worth our while to indulge in lengthy comments upon the mode of conducting this course of lectures, nor upon the success which has crowned the labors of its managers; but that the readers of this volume may know how truly national an interest had been excited in it, and for the sake of the future historian of American agricultural education, I will state that there have been registered on the book about 350 names. Of these persons, 172 only are from Connecticut, 23 from Massachusetts, 35 from New York, and the remainder is divided between Indiana, Kentucky, Vermont, Ohio, Pennsylvania, New Jersey, New Hampshire, Maine, Illinois, Florida, Wisconsin, Rhode Island, and the Canadas, East and West. Considering that in the Undergraduate Department of Yale there are only 502 students, the regular attendance of nearly or quite 350 at the agricultural lectures should be well weighed in the minds of the Faculty, and prompt them to not only give a tacit recognition, but, so far as consistent with professional duties, take an active interest in the establishment of this department of agriculture. They may rest assured that, by so doing, they will make the name of Yale more respected in her old age than it ever has been in her palmiest days. When, some time ago, I wanted to take a course of agricultural instruction, I was forced to cross the ocean, because there was no suitable place at home. Thanks to Prof. Porter, his associates, and the generous friends who have contributed their money to aid them, others will not be put to the same straits. Within a few years from this time, Yale College will have, probably, as spacious apartments, as complete a museum, library and reading-room, and as well appointed a laboratory as any student, however diligent, may require. And with this pleasing prospect in view, congratulating Profs. Porter and Johnson upon the success of their experiment, I close my note-book, and write among the things of the Past this first course of the YALE AGRICULTURAL LECTURES.

APPENDIX.

AN AGRICULTURAL EXPERIMENT.

NEAR the close of the Yale Agricultural Convention for 1860, Professor Porter promised to issue a scheme for some simple and easily-conducted experiments, with the results of which the Convention might contribute new material to the practice and theory of rational agriculture. The business of preparing a plan of trials having been confided to the undersigned, he has deemed it best to select some fertilizer as the subject of experiment, and indeed, that substance which in our country is everywhere accessible and cheap, its use being unhampered by the burdensome imposts which still render it expensive in nearly all other countries. This substance is salt; one which, furnishing indispensable ingredients to the digestive fluids, performs most important offices in the economy of the animal kingdom, and is, unquestionably, most naturally and healthfully derived from the food itself.

SAMUEL W. JOHNSON.

Yale Scientific School, New Haven, Ct., March, 1860.

EXPERIMENTAL STUDY OF THE USE OF SALT AS A FERTILIZER.

THE action of salt as a fertilizer, has long been a matter of uncertainty and dispute among agriculturists. In many cases it has been reported to be extremely useful, in many more to be entirely valueless, and in some positively disastrous.

We have no reason to disbelieve the testimony that has been offered at various times, and from a wide range of experimenters, although it is so contradictory in its character.

If the various statements concerning the use of salt as a fertilizer are true, the important question arises, How are we to know when it will be useful, and when otherwise?

This question can only be answered by the repetition of experiments, which must be made under a great variety of circumstances, and under conditions that are accurately known and defined.

In conducting such an inquiry, it is of the first importance to gather from the existing stock of experience, all the facts which throw any light either upon the question itself, or upon the methods of investigating it.

Under the conviction that a multitude of careful trials may be instituted among our farmers, with the prospect of explaining the contradictions of former experience, or at least of revealing the valuable fact that salt is capable of doing the agriculturist great service in many localities where it has not yet been tried, and also of contributing to the education of the public in the objects and methods of experimental agriculture, we have drawn up from various sources the facts, assertions and probabilities which may serve as guides in attempting the solution of this problem.

1*st.* We know that the constituents of common salt (chlorine and sodium) are unfailing ingredients of all agricultural plants, although they exist in vegetation in very variable, usually quite small amount.

2*d.* We know that in many instances (perhaps in all where this subject has been accurately studied) the use of salt as a manure has increased (often doubled) the amount of salt in the crop.

3*d.* We know that crops having large foliage contain (and require?) more salt than those of the small-leaved and few-leaved kinds.

4th. It is said that tobacco is largely increased in quantity, but injured in quality, by applying salt as a manure. The same is said of sugar-plants.

5th. It is probable that the white beet, mangel-wurzel, and carrot, among field-crops, (as is certain of asparagus in the garden,) being originally marine plants, will be more strikingly benefited by salt than other crops, and will admit of larger applications, other things being equal.

6th. We know that many soils near saline springs, (or reclaimed from salt marshes,) naturally contain as much or more salt than is needful for the growth of agricultural plants.

7th. We know that in many regions (those exposed to prevailing and especially stormy winds from the ocean) the soil annually receives from spray and rain more salt than is annually removed by crops.

8th. We know that salt is most often injurious in dry seasons, or on dry soils.

9th. It is probable that the positively injurious effects of salt are chiefly due to its being applied in too large quantity; for

10th. We know that a strong solution of salt hinders the germination of seeds, and destroys the life of the growing plant (marine plants of course excepted).

11th. We know (from the recent experiments of Sachs and Knop in Saxony) that a *weak* solution of salt hinders (by one half or more) the transpiration of water through the plant; therefore,

12th. It is probable that a little salt has the effect to keep the soil more humid, and thus tends to counteract drought; and,

13th. It is probable that a little salt, by hindering excessive transpiration, (and too rapid growth?) causes the cellular tissue of the plant to develop in a firmer, healthier manner than it might otherwise do; and thus may be explained,

14th. The assertion that a bushel or two of salt per acre on grain crops prevents falling (laying or lodging) of the straw.

15*th.* It is, however, the experience of Girardin, Fauchet, and Dubreuil, that large doses (more than 370 lbs. per acre) increase the straw rather than the grain, and make the crop lodge on soil that has been dunged.

16th. It is said that the small applications of salt make the straw of the grains brighter, and prevent rust.

17*th.* It is said that large applications delay the ripening of the grain.

18*th.* It is said that salt prevents potato rot (by delaying the sprouting and blossoming of the plant, so that the critical period of its life is brought after the hot fogs and rains of late summer?).

19*th.* We know, from many trials, (those of Kuhlmann, and recent ones of Liebig,) that salt often remarkably heightens the effect of other powerful manures.

20*th.* We know (from the studies of Way and Eichhorn) that salt is able to displace potash, ammonia, and lime from insoluble combinations of these bodies,—combinations such as, in all probability, exist in the soil. Therefore, and because

21*st.* We know that salt increases the power of water to dissolve the phosphates of lime, magnesia, &c.,

22*d.* It is probable that its use may, on certain soils, be equivalent to an application of these bodies, by rendering the stores of them already existing in the soil available to crops.

23*d.* It is probable that salt is sometimes advantageous, not so much as a fertilizer, as by destroying worms and the larvæ of insects.

24*th.* It is certain that fields well manured with stable or yard manure, made from cattle that are supplied with all the salt they desire, thus receive more salt than is removed from them in ordinary culture.

25*th.* It is probable that thorough-drained fields will be more benefited by (and require more?) salt, than undrained fields of similar soil.

26*th.* It is a matter of experience, that while 500 to 600, or even 800 lbs. of salt may be applied per acre before the seed, without injury (in moist climate or wet season), not more than 200 lbs. per acre should be put directly on the growing crop.

Any one may easily select for himself from the foregoing some one or more points that it is desirable to test in his own locality, and will also readily gather the most important circumstances that need to be regarded in carrying out an experiment to a good result.

We add, however, the following suggestions as to the manner of making the experiment:

I. Every experiment should furnish means of comparison with some standard. If, for example, it is sought to ascertain whether salt increases a crop on a given soil, not only should a portion of the crop and soil have salt applied to it, but another portion should be left without the application. If the question is, Is the straw strengthened, or the grain made heavier? then, obviously, opportunity must be given to observe how strong the straw is, or how heavy the grain is where no salt has been used.

II. The plots of ground should not usually consist in a strip a few feet wide, or in a few rows of the crop, but in a nearly square surface, so as to have as little *edge* to the piece as possible, for the roots of plants often extend several feet beyond ordinary dividing lines, if the soil be grateful to them.

III. The experimental ground should be as uniform as possible in quality of soil, in tillage, dunging, and exposure, and should all have had the same treatment as regards cropping and manuring for several years previous to the trial.

IV. The plots should be of good size, at least one-eighth, preferably one-fourth of an acre.

V. "Everything should be done by weight and measure;" guesswork is worse than useless. Let the plots be accurately measured, not "paced off." Let the materials added, and the crop removed, be carefully weighed, and not "estimated by the eye."

VI. Every care should be used to observe and record, with fulness and accuracy, the character, exposure, present condition and previous management of the soil. The climate and weather, the development of the crop in all its parts, and in all stages of its growth, and generally, all facts bearing on the experiment, should be taken into the account.

SCHEME OF EXPERIMENTS.

A. *General Effects of Salt*,—as increase of product, improvement of quality of crop, prevention of disease, &c.

Two plots of any soil in any crop,—both may receive other manures or not; but their treatment should differ *only* in this fact, that one is salted, the other not. Use the salt at the rate of 350 lbs. per acre, (see 26th observation).

B. *Effect on particular crops, or classes of crops*, as potatoes compared with carrots, grasses *vs.* root-crops, root-crops *vs.* grains.

Two plots for each crop as under A.

C. *Effects of different doses:*

Soil and crop alike,—one plot unsalted, one with 75 lbs, one with 150 lbs., one with 300 lbs., one with 450 lbs., or other different quantities, less or more in number, as convenient.

D. *Effects on different soils:*

Soils different,—tillage, manure, and crop the same. Dose of salt the same. Of each soil a salted and an unsalted plot should be observed.

CATALOGUE OF BOOKS

ON

AGRICULTURE AND HORTICULTURE,

PUBLISHED BY

C. M. SAXTON, BARKER & CO.,

No. 25 PARK ROW, NEW YORK.

SUITABLE FOR

SCHOOL, TOWN, AGRICULTURAL, & PRIVATE LIBRARIES.

AMERICAN FARMER'S ENCYCLOPEDIA, - - - - - - - **$4 00**

AS A BOOK OF REFERENCE FOR THE FARMER OR GARDENER, THIS Work is superior to any other. It contains Reliable Information for the Cultivation of every variety of Field and Garden Crops, the use of all kinds of Manures, descriptions and figures of American insects; and is, indeed, an Agricultural Library in itself, containing *twelve hundred pages*, octavo, and is illustrated by numerous engravings of Grasses, Grains, Animals, Implements, Insects, &c., &c. By GOUVERNEUR EMERSON OF PENNSYLVANIA.

AMERICAN WEEDS AND USEFUL PLANTS, - - - - **1 50**

AN ILLUSTRATED EDITION OF AGRICULTURAL BOTANY; An Enumeration and Description of Weeds and Useful Plants which merit the notice or require the attention of American Agriculturists. By WM. DARLINGTON, M. D. Revised, with Additions, by GEORGE THURBER, Prof. of Mat. Med. and Botany in the New York College of Pharmacy. Illustrated with nearly 300 Figures, drawn expressly for this work.

ALLEN'S (R. L.) AMERICAN FARM BOOK, - - - - - - **1 00**

OR A COMPEND OF AMERICAN AGRICULTURE; being a Practical Treatise on Soils, Manures, Draining, Irrigation, Grasses, Grain, Roots, Fruits, Cotton, Tobacco, Sugar Cane, Rice, and every Staple Product of the United States; with the best methods of Planting, Cultivating and Preparation for Market. Illustrated with more than 100 engravings.

ALLEN'S (R. L.) DISEASES OF DOMESTIC ANIMALS, - - **75**

BEING A HISTORY AND DESCRIPTION OF THE HORSE, MULE, CATTLE, Sheep, Swine, Poultry and Farm Dogs, with Directions for their Management, Breeding, Crossing, Rearing, Feeding, and Preparation for a Profitable Market; also, their Diseases and Remedies, together with full Directions for the Management of the Dairy, and the comparative Economy and Advantages of Working Animals,—the Horse, Mule, Oxen, &c.

ALLEN'S (L. F.) RURAL ARCHITECTURE, - - - - - - **1 25**

BEING A COMPLETE DESCRIPTION OF FARM HOUSES, COTTAGES AND Out Buildings, comprising Wood Houses, Workshops, Tool Houses, Carriage and Wagon Houses, Stables, Smoke and Ash Houses, Ice Houses, Apiaries or Bee Houses, Poultry Houses, Rabbitry, Dovecote, Piggery, Barns and Sheds for Cattle, &c., &c.; together with Lawns, Pleasure Grounds and Parks; the Flower, Fruit and Vegetable Garden; also, the best method of conducting water into Cattle Yards and Houses. Beautifully illustrated.

ALLEN (J. FISK) ON THE CULTURE OF THE GRAPE, - - **1 00**

A PRACTICAL TREATISE ON THE CULTURE AND TREATMENT OF THE Grape Vine, embracing its History, with Directions for its Treatment in the United States of America, in the Open Air and under Glass Structures, with and without Artificial Heat

Mailed post paid upon receipt of price.

AMERICAN ARCHITECT, - - - - - - - - - - - - 6 00

Comprising Original Designs of Cheap Country and Village Residences, with Details, Specifications, Plans and Directions, and an Estimate of the Cost of each Design. By John W. Ritch, Architect. First and Second Series, 4to, bound in 1 vol.

AMERICAN FLORIST'S GUIDE, - - - - - - - - - - 75

Comprising the American Rose Culturist, and Every Lady her own Flower Gardener.

BARRY'S FRUIT GARDEN, - - - - - - - - - - - 1 25

A Treatise, Intended to Explain and Illustrate the Physiology of Fruit Trees, the Theory and Practice of all Operations connected with the Propagation, Transplanting, Pruning and Training of Orchard and Garden Trees, as Standards, Dwarfs, Pyramids, Espalier, &c. The Laying out and Arranging different kinds of Orchards and Gardens, the selection of suitable varieties for different purposes and localities, Gathering and Preserving Fruits, Treatment of Diseases, Destruction of Insects, Description and Uses of Implements, &c. Illustrated with upwards of 150 Figures. By P. Barry, of the Mount Hope Nurseries, Rochester, N. Y.

BEMENT'S (C. N.) RABBIT FANCIER, - - - - - - - - 50

A Treatise on the Breeding, Rearing, Feeding and General Management of Rabbits, with Remarks upon their Diseases and Remedies, to which are added Full Directions for the Construction of Hutches, Rabbitries, &c., together with Recipes for Cooking and Dressing for the Table. Beautifully illustrated.

BLAKE'S (REV. JOHN L.) FARMER AT HOME, - - - - - 1 25

A Family Text Book for the Country; being a Cyclopedia of Agricultural Implements and Productions, and of the more important topics in Domestic Economy, Science and Literature, adapted to Rural Life. By Rev. John L. Blake, D. D.

BOUSSINGAULT'S (J. B.) RURAL ECONOMY, - - - - - - 1 25

Or, Chemistry Applied to Agriculture; presenting Distinctly and in a Simple Manner the Principles of Farm Management, the Preservation and Use of Manures, the Nutrition and Food of Animals, and the General Economy of Agriculture. The work is the fruit of a long life of study and experiment, and its perusal will aid the farmer greatly in obtaining a practical and scientific knowledge of his profession.

BROWNE'S AMERICAN BIRD FANCIER, - - - - - - - 25

The Breeding, Rearing, Feeding, Management and Peculiarities of Cage and House Birds. Illustrated with engravings.

BROWNE'S AMERICAN POULTRY YARD, - - - - - - - 1 00

Comprising the Origin, History and Description of the Different Breeds of Domestic Poultry, with Complete Directions for their Breeding, Crossing, Rearing, Fattening and Preparation for Market; including specific directions for Caponizing Fowls, and for the Treatment of the Principal Diseases to which they are subject, drawn from authentic sources and personal observation. Illustrated with numerous engravings.

BROWNE'S (D. JAY) FIELD BOOK OF MANURES, - - - - - 1 25

Or, American Muck Book; Treating of the Nature, Properties, Sources, History and Operations of all the Principal Fertilizers and Manures in Common Use, with specific directions for their Preservation and Application to the Soil and to Crops; drawn from authentic sources, actual experience and personal observation, as combined with the Leading Principles of Practical and Scientific Agriculture.

BRIDGEMAN'S (THOS.) YOUNG GARDENER'S ASSISTANT, - - 1 50

In Three Parts; Containing Catalogues of Garden and Flower Seed, with Practical Directions under each head for the Cultivation of Culinary Vegetables, Flowers, Fruit Trees, the Grape Vine, &c.; to which is added a Calendar to each part, showing the work necessary to be done in the various departments each month of the year. One volume octavo.

BRIDGEMAN'S KITCHEN GARDENER'S INSTRUCTOR, ½ Cloth, 50
" " " " Cloth, 60

Mailed post paid upon receipt of price.

BRIDGEMAN'S FLORIST'S GUIDE, - - - - - - - ½ Cloth, 50
" " " - - - - - - Cloth, 60

BRIDGEMAN'S FRUIT CULTIVATOR'S MANUAL, - ½ Cloth, 50
" " " " - - Cloth, 60

BRECK'S BOOK OF FLOWERS, - - - - - - - - - - - 1 00

In which are Described all the Various Hardy Herbaceous Perennials, Annuals, Shrubs, Plants and Evergreen Trees, with Directions for their Cultivation.

BUIST'S (ROBERT) AMERICAN FLOWER GARDEN DIRECTORY, 1 25

Containing Practical Directions for the Culture of Plants, in the Flower Garden, Hothouse, Greenhouse, Rooms or Parlor Windows, for every month in the Year; with a Description of the Plants most desirable in each, the nature of the Soil and situation best adapted to their Growth, the Proper Season for Transplanting, &c.; with Instructions for erecting a Hothouse, Greenhouse, and Laying out a Flower Garden; the whole adapted to either Large or Small Gardens, with Instructions for Preparing the Soil, Propagating, Planting, Pruning, Training and Fruiting the Grape Vine.

BUIST'S (ROBERT) FAMILY KITCHEN GARDENER, - - - 75

Containing Plain and Accurate Descriptions of all the Different Species and Varieties of Culinary Vegetables, with their Botanical, English, French and German names, alphabetically arranged, with the Best Mode of Cultivating them in the Garden or under Glass; also Descriptions and Character of the most Select Fruits, their Management, Propagation, &c. By Robert Buist, author of the "American Flower Garden Directory," &c.

CHINESE SUGAR CANE AND SUGAR-MAKING, - - - - - 25

Its History, Culture and Adaptation to the Soil, Climate and Economy of the United States, with an Account of Various Processes of Manufacturing Sugar. Drawn from authentic sources, by Charles F. Stansbury, A. M., late Commissioner at the Exhibition of all Nations at London.

CHORLTON'S GRAPE-GROWER'S GUIDE, - - - - - - - 60

Intended Especially for the American Climate. Being a Practical Treatise on the Cultivation of the Grape Vine in each department of Hothouse, Cold Grapery, Retarding House and Out-door Culture. With Plans for the construction of the Requisite Buildings, and giving the best methods for Heating the same. Every department being fully illustrated. By William Chorlton.

COBBETT'S AMERICAN GARDENER, - - - - - - - - - 50

A Treatise on the Situation, Soil and Laying-out of Gardens, and the Making and Managing of Hotbeds and Greenhouses, and on the Propagation and Cultivation of the several sorts of Vegetables, Herbs, Fruits and Flowers.

COTTAGE AND FARM BEE-KEEPER, - - - - - - - - - 50

A Practical Work, by a Country Curate.

COLE'S AMERICAN FRUIT BOOK, - - - - - - - - - - 50

Containing Directions for Raising, Propagating and Managing Fruit Trees, Shrubs and Plants; with a Description of the Best Varieties of Fruit, including New and Valuable Kinds.

COLE'S AMERICAN VETERINARIAN, - - - - - - - - 50

Containing Diseases of Domestic Animals, their Causes, Symptoms and Remedies; with Rules for Restoring and Preserving Health by good management; also for Training and Breeding.

DADD'S AMERICAN CATTLE DOCTOR, - - - - - - - - - 1 00

Containing the Necessary Information for Preserving the Health and Curing the Diseases of Oxen, Cows, Sheep and Swine, with a Great Variety of Original Recipes and Valuable Information in reference to Farm and Dairy Management, whereby every Man can be his own Cattle Doctor. The principles taught in this work are, that all Medication shall be subservient to Nature—that all Medicines must be sanative in their operation, and administered with a view of aiding the vital powers, instead of depressing, as heretofore, with the lancet or by poison. By G. H. Dadd, M. D. Veterinary practitioner.

Mailed post paid upon receipt of price.

DADD'S MODERN HORSE DOCTOR, - - - - - - - - - 1 00

An American Book for American Farmers; Containing Practical Observations on the Causes, Nature and Treatment of Disease and Lameness of Horses, embracing the Most Recent and Approved Methods, according to an enlightened system of Veterinary Practice, for the Preservation and Restoration of Health. With illustrations.

DADD'S ANATOMY AND PHYSIOLOGY OF THE HORSE, Plain, - 2 00
" " " " " Colored Plates, 4 00

With Anatomical and Questional Illustrations; Containing, also, a Series of Examinations on Equine Anatomy and Philosophy, with Instructions in reference to Dissection and the mode of making Anatomical Preparations; to which is added a Glossary of Veterinary Technicalities, Toxicological Chart, and Dictionary of Veterinary Science.

DANA'S MUCK MANUAL, FOR THE USE OF FARMERS, - - 1 00

A Treatise on the Physical and Chemical Properties of Soils and Chemistry of Manures; including, also, the subject of Composts, Artificial Manures and Irrigation. A new edition, with a Chapter on Bones and Superphosphates.

DANA'S PRIZE ESSAY ON MANURES, - - - - - - - 25

Submitted to the Trustees of the Massachusetts Society for Promoting Agriculture, for their Premium. By Samuel H. Dana.

DOMESTIC AND ORNAMENTAL POULTRY, Plain Plates, - - - 1 00
" " " Colored Plates, - - 2 00

A Treatise on the History and Management of Ornamental and Domestic Poultry. By Rev. Edmund Saul Dixon, A. M., with large additions by J. J. Kerr, M. D. Illustrated with sixty-five Original Portraits, engraved expressly for this work. Fourth edition, revised.

DOWNING'S (A. J.) LANDSCAPE GARDENING, - - - - - - 3 50

Revised, Enlarged and Newly Illustrated, by Henry Winthrop Sargent. This Great Work, which has accomplished so much in elevating the American Taste for Rural Improvements, is now rendered doubly interesting and valuable by the experience of all the Prominent Cultivators of Ornamental Trees in the United States, and by the descriptions of American Places, Private Residences, Central Park, New York, Llewellyn Park, New Jersey, and a full account of the Newer Deciduous and Evergreen Trees and Shrubs. The illustrations of this edition consist of *seven superb steel plate engravings*, by Smillie, Hinshelwood, Duthie and others; besides *one hundred engravings on wood and stone*, of the best American Residences and Parks, with Portraits of many New or Remarkable Trees and Shrubs.

DOWNING'S (A. J.) RURAL ESSAYS, - - - - - - - - - 3 00

On Horticulture, Landscape Gardening, Rural Architecture, Trees, Agriculture, Fruit, with his Letters from England. Edited, with a Memoir of the Author, by George Wm. Curtis, and a Letter to his Friends, by Frederika Bremer, and an elegant Steel Portrait of the Author.

EASTWOOD (B.) ON THE CULTIVATION OF THE CRANBERRY, 50

With a Description of the Best Varieties. By B. Eastwood, "Septimus," of the New York Tribune. Illustrated.

ELLIOTT'S WESTERN FRUIT BOOK, - - - - - - - - - 1 25

A New Edition of this Work, Thoroughly Revised. Embracing all the New and Valuable Fruits, with the Latest Improvements in their Cultivation, up to January, 1859. especially adapted to the wants of Western Fruit Growers; full of excellent illustrations. By F. R. Elliott, Pomologist, late of Cleveland, Ohio, now of St. Louis.

EVERY LADY HER OWN FLOWER GARDENER, - - - - 50

Addressed to the Industrious and Economical only; containing simple and practical Directions for Cultivating Plants and Flowers; also, Hints for the Management of Flowers in Rooms, with brief Botanical Descriptions of Plants and Flowers. The whole in plain and simple language. By Louisa Johnson.

Mailed post paid upon receipt of price.

FARM DRAINAGE, - - - - - - - - - - - - - - 1 00

The Principles, Processes and Effects of Draining Land, with Stones, Wood, Drain-plows, Open Ditches, and especially with Tiles; including Tables of Rainfall, Evaporation, Filtration, Excavation, capacity of Pipes, cost and number to the acre. With more than 100 illustrations. By the Hon. Henry F. French, of New Hampshire.

FESSENDEN'S (T. G.) AMERICAN KITCHEN GARDENER, - - 50

Containing Directions for the Cultivation of Vegetables and Garden Fruits. Cloth.

FESSENDEN'S COMPLETE FARMER AND AMERICAN GARDENER, 1 25

Rural Economist and New American Gardener; Containing a Compendious Epitome of the most Important Branches of Agriculture and Rural Economy; with Practical Directions on the Cultivation of Fruits and Vegetables, including Landscape and Ornamental Gardening. By Thomas G. Fessenden. 2 vols. in 1.

FIELD'S PEAR CULTURE, - - - - - - - - - - - - - 1 00

The Pear Garden; or, a Treatise on the Propagation and Cultivation of the Pear Tree, with Instructions for its Management from the Seedling to the Bearing Tree. By Thomas W. Field.

FISH CULTURE, - - - - - - - - - - - - - - 1 00

A Treatise on the Artificial Propagation of Fish, and the Construction of Ponds, with the Description and Habits of such kinds of Fish as are most suitable for Pisciculture. By Theodatus Garlick, M. D., Vice-President of the Cleveland Academy of Nat. Science.

FLINT ON GRASSES, - - - - - - - - - - - - - 1 25

A Practical Treatise on Grasses and Forage Plants; Comprising their Natural History, Comparative Nutritive Value, Methods of Cultivation, Cutting, Curing and the Management of Grass Lands. By Charles L. Flint, A. M., Secretary of the Mass. State Board of Agriculture.

GUENON ON MILCH COWS, - - - - - - - - - - 60

A Treatise on Milch Cows, whereby the Quality and Quantity of Milk which any Cow will give may be accurately determined by observing Natural Marks or External Indications alone; the length of time she will continue to give Milk, &c., &c. By M. Francis Guenon, of Libourne, France. Translated by Nicholas P. Trist, Esq.; with Introduction, Remarks and Observations on the Cow and the Dairy, by John S. Skinner. Illustrated with numerous Engravings. Neatly done up in paper covers, 37 cts.

HERBERT'S HINTS TO HORSE-KEEPERS, - - - - - - 1 25

Complete Manual for Horsemen; Embracing:

How to Breed a Horse.
How to Buy a Horse.
How to Break a Horse.
How to Use a Horse.
How to Feed a Horse.
How to Physic a Horse.
(Allopathy and Homœopathy
How to Groom a Horse.
How to Drive a Horse.
How to Ride a Horse.

And Chapters on Mules and Ponies. By the late Henry William Herbert (Frank Forrester); with additions, including Rarey's Method of Horse Taming, and Baucher's System of Horsemanship; also, giving directions for the Selection and Care of Carriages and Harness of every description, from the City "Turn Out" to the Farmer's "Gear," and a Biography of the eccentric Author. *Illustrated throughout.*

HOOPER'S DOG AND GUN, - - - - - - - - - - 50

A Few Loose Chapters on Shooting, among which will be found some Anecdotes and Incidents; also Instructions for Dog Breaking, and interesting letters from Sportsmen. By A Bad Shot.

HYDE'S CHINESE SUGAR CANE, - - - - - - - - - 25

Containing its History, Mode of Culture, Manufacture of the Sugar, &c.; with Reports of its success in different parts of the United States.

Mailed post paid upon receipt of price.

JOHNSTON'S (JAMES F. W.) AGRICULTURAL CHEMISTRY, - - 1 25

Lectures on the Application of Chemistry and Geology to Agriculture. New Edition, with an Appendix, containing the Author's Experiments in Practical Agriculture.

JOHNSTON'S (J. F. W.) ELEMENTS OF AGRICULTURAL CHEMISTRY AND GEOLOGY, - - - - - - - - - - 1 00

With a Complete Analytical and Alphabetical Index, and an American Preface. By Hon. Simon Brown, Editor of the "New England Farmer."

OHNSTON'S (J. F. W.) CATECHISM OF AGRICULTURAL CHEMISTRY AND GEOLOGY, - - - - - - - - - - 25

By James F. W. Johnston, Honorary Member of the Royal Agricultural Society of England, and author of "Lectures on Agricultural Chemistry and Geology." With an Introduction by John Pitkin Norton, M. A., late Professor of Scientific Agriculture in Yale College. With Notes and Additions by the Author, prepared expressly for this edition, and an Appendix compiled by the Superintendent of Education in Nova Scotia. Adapted to the use of Schools.

LANGSTROTH (REV. L. L.) ON THE HIVE AND HONEY BEE, - 1 25

A Practical Treatise on the Hive and Honey Bee, Third edition, enlarged and *illustrated with numerous engravings.* This Work is, without a doubt, the best work on the Bee published in any language, whether we consider its scientific accuracy, the practical instructions it contains, or the beauty and completeness of its illustrations.

LEUCHARS' HOW TO BUILD AND VENTILATE HOTHOUSES, - 1 25

A Practical Treatise on the Construction, Heating and Ventilation of Hothouses, including Conservatories, Greenhouses, Graperies and other kinds of Horticultural Structures; with Practical Directions for their Management, in regard to Light, Heat and Air. Illustrated with numerous engravings. By P. B. Leuchars, Garden Architect.

LIEBIG'S (JUSTUS) FAMILIAR LECTURES ON CHEMISTRY, - 50

And its relation to Commerce, Physiology, and Agriculture. Edited by John Gardener, M. D.,

LINSLEY'S MORGAN HORSES, - - - - - - - - 1 00

A Premium Essay on the Origin, History, and Characteristics of this remarkable American Breed of Horses; tracing the Pedigree from the original Justin Morgan, through the most noted of his progeny, down to the present time. With numerous portraits. To which are added Hints for Breeding, Breaking and General Use and Management of Horses, with practical Directions for Training them for Exhibition at Agricultural Fairs. By D. C. Linsley, Editor of the American Stock Journal.

MOORE'S RURAL HAND BOOKS, - - - - - - - - - - 1 25

First Series, containing Treatises on—

The Horse,	The Pests of the Farm,
The Hog,	Domestic Fowls, and
The Honey Bee,	The Cow.

Second Series, containing— - - - - - 1 25

Every Lady her own Flower Gardener,	Essay on Manures,
Elements of Agriculture,	American Kitchen Gardener,
Bird Fancier,	American Rose Culturist.

Third Series, containing— - - - - - - 1 25

Miles on the Horse's Foot,	Vine-Dresser's Manual,
The Rabbit Fancier,	Bee-Keeper's Chart,
Weeks on Bees,	Chemistry Made Easy.

Fourth Series, containing— - - - - - 1 25

Persoz on the Vine,	Hooper's Dog and Gun,
Liebig's Familiar Letters,	Skillful Housewife,

Browne's Memoirs of Indian Corn.

Mailed post paid upon receipt of price.

www.ingramcontent.com/pod-product-compliance
Lightning Source LLC
LaVergne TN
LVHW011219110826
845150LV00006B/1482
9781425516147